THE SNAILBEACH DISTRICT RAILWAYS

No.2 on train at Pontesbury with fitter-driver Dickie Preece and another workman, early 1940s. The driver is holding a long oiler

THE SNAILBEACH DISTRICT RAILWAYS

With introductions to Snailbeach mine
and the railway schemes of South West Shropshire

BY
ANDY CUCKSON

TWELVEHEADS PRESS

TRURO 2017

CONTENTS

Front cover main picture:

No.2 outside the locomotive shed in 1943, with the Baldwin locomotives *No.3* and *No.4* inside.
L. W. PERKINS

Bottom left to right:

No.4 at Snailbeach c.1925.
COLLECTION P. INGHAM

Fernhill on the long siding with crew and miners at Snailbeach mine.
NATIONAL RAILWAY MUSEUM/SCIENCE & SOCIETY PICTURE LIBRARY

The W. G. Bagnall works photograph of *Dennis*.
COURTESY D. A. BREWER

Back cover:

Fitter-driver David Reese Jones on board *No.4* at Pontesbury. J. F. BRUTON, NATIONAL RAILWAY MUSEUM/SCIENCE & SOCIETY PICTURE LIBRARY

TWELVEHEADS PRESS

First published in 2017 by Twelveheads Press
Second edition 2017 by Twelveheads Press
Truro, Cornwall.
ISBN 978 0 906294 932
British Library Cataloguing-in-Publication Data.
A catalogue record for this book is available from the British Library.
Typeset in Baskerville and Helvetica
Printed by Short Run Press Ltd, Exeter

FOREWORD

By Michael Iggulden
Chairman of the Snailbeach District Railways Company

Reading the interesting details of the SDRC has brought back many memories. I particularly recall the times when the Directors carried out their annual inspection of the line. It was then that the residents of Snailbeach were on their best behaviour as 'the Men from the Railway' were in the village. The villagers made sure there were no cars parked on the company's land and there were no signs of trespass. I feel sure that all went back to normal once we had left!

Andy Cuckson is to be congratulated on the considerable research he has undertaken in preparing this account of the Snailbeach District Railways Company. I am grateful to him for bringing all the threads together and for presenting the complete picture that is the history of the line.

This book is also a welcome addition to the history of the life of Colonel Stephens, with his dedication to light railways in the early part of the last century. I recommend it to all those interested in the early workings of railways in this country who will surely enjoy learning more of this little known line.

M. J. L. Iggulden FRICS
Chairman – SDRC

Common seal of the Snailbeach District Railways Company.
The four lobed device in the centre is a monogram of the letters T-S-D-R-C-o.
COURTESY M. J. IGGULDEN

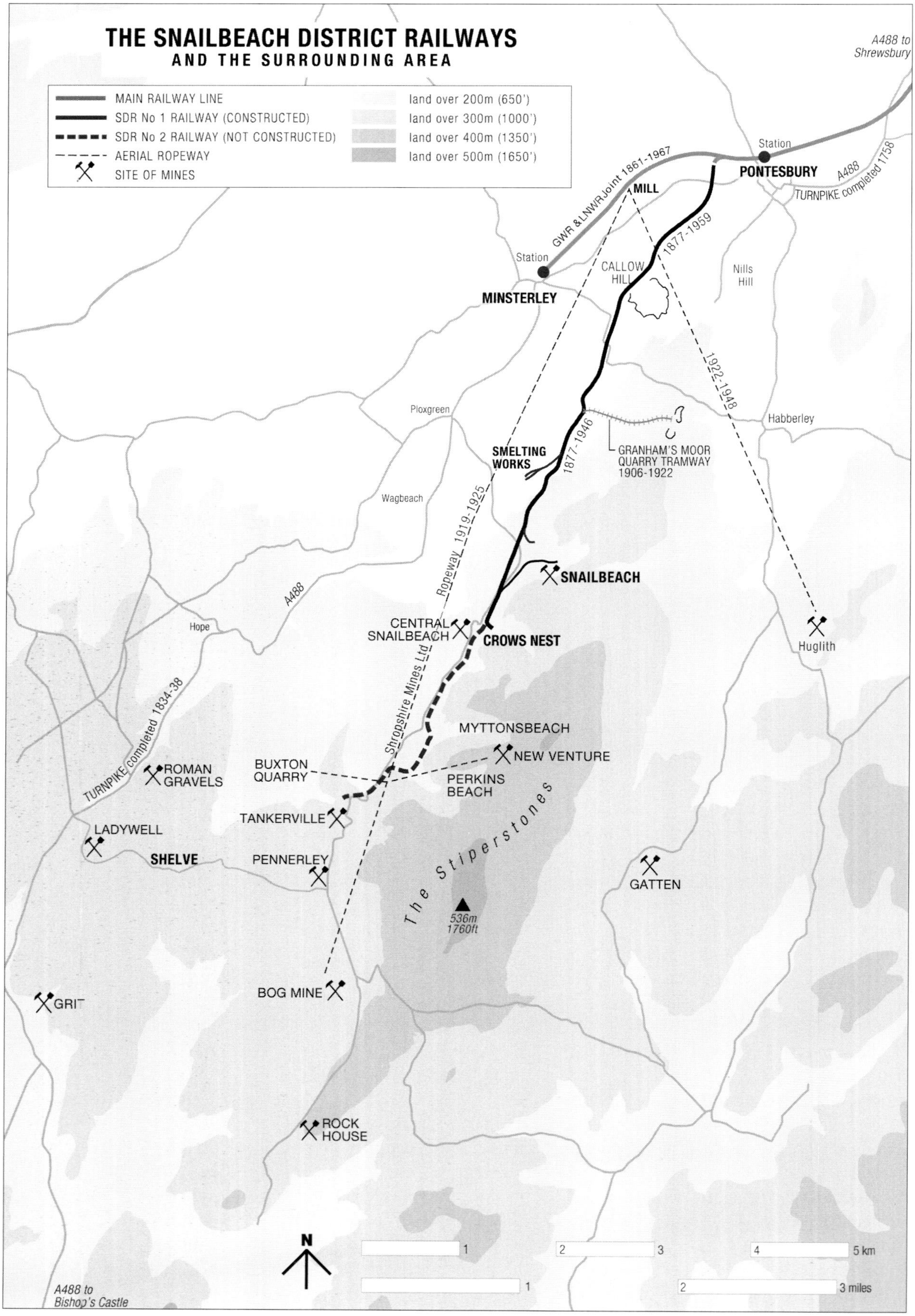
THE SNAILBEACH DISTRICT RAILWAYS
AND THE SURROUNDING AREA
MAIN RAILWAY LINE
SDR No 1 RAILWAY (CONSTRUCTED)
SDR No 2 RAILWAY (NOT CONSTRUCTED)
AERIAL ROPEWAY
SITE OF MINES
land over 200m (650')
land over 300m (1000')
land over 400m (1350')
land over 500m (1650')
A488 to Shrewsbury
Station
PONTESBURY
A488 TURNPIKE completed 1758
GWR & LNWR Joint 1861-1967
MILL
1877-1959
Station
MINSTERLEY
CALLOW HILL
Nills Hill
1922-1948
Habberley
Ploxgreen
SMELTING WORKS
1877-1946
GRANHAM'S MOOR QUARRY TRAMWAY 1906-1922
Wagbeach
Ropeway 1919-1925
SNAILBEACH
A488
Hope
CENTRAL SNAILBEACH
CROWS NEST
Huglith
Shropshire Mines Ltd
TURNPIKE completed 1834-38
MYTTONSBEACH
NEW VENTURE
ROMAN GRAVELS
BUXTON QUARRY
PERKINS BEACH
TANKERVILLE
LADYWELL
SHELVE
PENNERLEY
The Stiperstones
GATTEN
536m 1760ft
GRIT
BOG MINE
ROCK HOUSE
N
5 km
3 miles
A488 to Bishop's Castle

PREFACE

The Snailbeach District Railways was one of a small number of railway building projects engineered by Henry Dennis. He was a Cornishman who became wealthy through personal enterprise and was a highly regarded captain of industry. Railway building was a small part of his career, but it was where he started. Dennis built the Snailbeach line to serve extractive industries in which his wide engineering knowledge brought great success. Its two original principal freight commodities were lead for sale, and coal for the mine engines and smelting furnaces.

In his Foreword, Michael Iggulden notes that this book extends the history of Holman Fred Stephens, the well-known engineer and promoter of minor railways, but in this case Stephens merely renovated a railway built by Dennis. To put Dennis's experience into perspective, his known railways are surveyed briefly in an early chapter. The lives, works and spheres of influence of both men are sketched later in this book, but Stephens' railway experience is covered well by other authors, and need not be repeated here.

This book really began when I was a boy of about thirteen. I had caught the model railway bug and wanted to learn about the real thing – especially narrow gauge railways, for some reason. In the local library, the David & Charles series *A Regional History of the Railways of Great Britain* was scoured for references to such lines. One of them had such an exotic-sounding name it almost couldn't be true: the Snailbeach District Railways. Interest was kindled, but the first problem was to find out exactly where Snailbeach was! The specialist library of the Narrow Gauge Railway Society answered this and some other questions. Over Whitsun 1971, I visited Snailbeach for the first time, exploring and recording as much as I could of the mine and the railway down to Pontesbury. Another twenty years passed before I began any serious research.

When I finally set out to write this book, I wanted it to be as readable and as interesting as possible to as many people as possible – even if they did not live in the locality of the subject and had little knowledge of or interest in railways. This is a tall order when most people, unlike railway enthusiasts, would not find a small, grimy mineral railway intrinsically interesting. To this end, I have aimed to describe in proper context the Snailbeach District Railways Company and the railway it built and operated; to explain in reasonable depth its origins, the turn of events and decision making, the various periods of advance and decline in trade, and the different functions and features of the line. But for many people there are more basic questions to be answered. The non-technical and non-specialist person can be put off by jargon and excessive detail, so I have tried to minimise such things. To keep this book accessible, there is a glossary of technical terms. The explanations might not cover every question, and in such cases I hope readers will be encouraged to look a little further elsewhere. I have excluded any terms which could be found directly or indirectly in the Concise Oxford Dictionary of Current English (Seventh edition).

Three authors originally inspired me to write this. By 1974, two editions of *The Snailbeach District Railways* by Eric S. Tonks had appeared, but even the revised edition seemed only to give a good picture of the line's later history, and not of the early years. This was part of the mystery of the line, and I was determined to clarify it. *The Shropshire Lead Mines* by Fred Brook and Martin Allbutt had also appeared by then, which threw a little more light on how in its early years the railway served Snailbeach mine. In reading these books I realised that the mining industry at Snailbeach also needed to be looked at closely, or the part played by the railway would not be fully understood. Thorough research of Snailbeach mining records, as well as the railway records, allowed a better portrayal of the railway's early times. They provided the necessary original 'what', 'why', 'how' and even 'who' of this small and rather enigmatic narrow gauge railway.

I have also been able to take a fresh look at the so-called 'Colonel Stephens era' (post-1922) of the Company. For this, local people who knew the railwaymen or were related to them have been an important source of eye-witness accounts. Occasional nuggets of information handed down by parents or grandparents have also been reported. I found that my informants were consistent and reliable (admitting when they were not certain), and often corroborated each other when spoken to individually. Sometimes they fleshed out information given in existing documents, and even mentioned new people and issues which are not known to be on record. This part of the work – certainly for me – breathed life into the book.

While researching the subject, the author has written and published a number of articles about the railway in magazines and society journals. Any errors found in those articles since publication have been corrected here, but there is also valid material in them not included in this book (most often to keep it brief) which might repay reading for the more serious student. They are noted in the references.

Although I hope that all readers will enjoy this book, it was always my special intention that the residents of Snailbeach and the district would have a scrupulously factual account of their old railway which they can trust. I believe that I have achieved this goal, but if there are any untoward mistakes, the fault is entirely mine.

UNITS OF MEASUREMENT

Units of measurement and money used in this book are those which were concurrent with events described. They may be converted as follows:

Money:	£1 =	20 shillings (s) = 100 pence (p)
		1 guinea = 21 shillings
		1 shilling = 12 pence (d)
Length:	1 mile =	8 furlongs = 80 chains = 1,760 yards = 1.6093 kilometres
		1 yard = 3 feet = 36 inches = 0.9144 metres
Area:	1 acre =	4 roods = 160 perches = 0.45 hectare
Weight	1 ton =	20 hundredweight (cwt) = 1.016 tonnes
Volume	1 gallon =	4.546 litres

ACKNOWLEDGEMENTS

The following local people either helped by providing information and photographs directly, or helped me to locate those who could. I am only sorry that a number of them are no longer with us to see the fruits of our labour now that this book is in print.

The late J. Butler; the late Mr D. Challinor and Mrs M. Challinor; the late R. G. Cooke and Mrs S. Cooke; Mrs E. Humphreys; the late G. France; P. Francis; the late Mrs G. (Dolly) Harrison; the late Mr and Mrs L. Hordley; Mrs N. Morris; M. and D. Owen; Mrs E. Rowson; the late Miss M. Rowson; the late Mrs N. Rowson; Mr and Mrs R. Rowson; Mrs S. Rowson; N. Southwick; Mrs M. Tate; and Mr and Mrs A. Yapp.

The late Mrs M. (Peggy) Chidley and her family, as well as supplying useful information, are also accorded special thanks for allowing the use of photographs which have been in the family for over a century, and are thought to have been taken by their uncle Albert Hewitt, who was a keen photographer. They are arguably the oldest yet found which show the Snailbeach line, and take us close to the year of opening. They are a marvellous find. The photograph of locomotive *Fernhill*, now held in the National Railway Museum, was originally also part of the family set.

This book has been written at the same time as a new book *Rails to Glyn Ceiriog – The History of the Glyn Valley Tramway* by John Milner and Beryl Williams (Ceiriog Press/Rail Romances). There are various overlapping areas in the history of both railways, and to ensure that good, factual works appear in print, John, Beryl and I have co-operated fully with a free exchange of information at all times while writing our respective books.

Other people who have helped with information and photographs are: A. Austin; A. C. Baker; A. Barrell; M. Bevan; V. J. Bradley; the late D. A. Brewer and Mrs H. Brewer; I. J. Brown; T. Burnham; I. Cooper; G. Fairhurst; M. C. Gill; S. Hargraves; P. G. Hindley; G. C. Hughes; M. J. L. Iggulden; P. Ingham; B. Janes; K. Lake; J. Lawson; S. A. Leleux; M. J. T. Lewis; R. C. Link; I. M. Lloyd; the late K. C. Lock; M. J. Messenger; the late E. J. R. Miller; the late C. Pealling; D. J. Pomeroy; D. Poyner; the late A. Pratt; K. P. Seaward; M. J. Shaw; P. Shaw; F. W. Shuttleworth; J. Smith; P. Strange; N. Taylor; P. Teather; G. Toms; the late J. G. Vincent; R. Wear; Mrs S. Whitehouse; and D. Woodcock.

By far the greatest volume of information and evidence has been supplied by the following companies, organisations and voluntary societies through the generous help of their officers, who for the most part I am unable to name individually:

The Birmingham Public Libraries; the British Library; the Colonel Stephens Railway Museum Archive; the Denbighshire Historical Society; the Denbighshire Record Office; the Flintshire Record Office; the Historical Model Railway Society; the Industrial Locomotive Society; the Industrial Railway Society; the Ironbridge Gorge Museum Trust; Land Registry Telford Office; the Llangollen Museum; Longueville Gittins; the Manchester Public Libraries; the Narrow Gauge Railway Society; the National Archives; the National Library of Wales; the National Railway Museum; the Parliamentary Archives; the Shrewsbury Chronicle; the Shropshire Archives and Local Studies Centre; the Shropshire Caving & Mining Club; Shropshire Registration Service; Terex Pegson Ltd; and Wrexham Archives.

Extracts from Ordnance Survey maps and plans, and drawings based on them, are reproduced with Crown Copyright reserved

I am deeply grateful to them all.

PUBLISHERS; NOTE

The publishers are grateful to David H. Smith for his hard work on the drawings of locomotives and wagons, and to Mrs Elaine Flint for permission to use the late E. A. Wade's wagon drawing.

A post card with three views of Snailbeach, one of Crowsnest in the centre and one of Stiperstones bottom left. Upper Works and the reversing siding feature in the top left view, while the SDR main line passes between Farm Cottage and Prospect House in the top right. The interior view of the Cosy Café (bottom right) is certainly post-Great War. The café was to be found in the old Nissen hut, long a fixture in the middle of Snailbeach. AUTHOR'S COLLECTION

CHAPTER ONE

Introduction

Some twelve miles to the south west of Shrewsbury in Shropshire, the village of Snailbeach nestles high on the west side of the Stiperstones ridge. It looks out over the picturesque Hope Valley, further west to the Long Mountain (the traditional border with Wales) and then to Wales itself. The name Snailbeach was originally the name of the hanging valley which cuts roughly east into the ridge. It later became the name of the nearest farm, and of the lead mine which was the cause of an expanding settlement.

Lead has been mined in south west Shropshire at least as far back as the Roman occupation. A pig of lead bearing the imprint of the Emperor Hadrian was found by workmen in south west Shropshire in the late eighteenth century.[1] In Medieval times, the Caus and Minsterley estate of the Earl of Stafford included a large part if not all of the Hogstow Forest. This straddled the Hope Valley, which was the main location of the south west Shropshire lead ore-field. In that period and through to the sixteenth century there are records of lead mining in the Hogstow Forest. Parts of Stafford's estate passed to Sir John Thynne as a marriage settlement (only later were the Thynnes Viscounts Weymouth and Marquises of Bath). Weymouth's Minsterley estate included a large area on the western flanks of the Stiperstones, with lead ore deposits at Callow Hill and Snailbeach.

The earliest firm references to mining at Snailbeach occur in the seventeenth and mid-eighteenth centuries, when there were brief spells of shallow, low cost mining with unspectacular output. High investment and deep exploration, which gave the mine a reputation for richness, began in the 1780s. Although Snailbeach Farm was established in the seventeenth century, more extensive habitation centred on Snailbeach mine only began at the end of the eighteenth century, with the formation by Thomas Lovett of Chirk of a private company of mine investors. This Snailbeach mine

A postcard view of Snailbeach post-1922. To the left a few of the mine buildings are seen, including the locomotive shed, which by now has the corrugated sheet storage stockade built onto its north side and the extension shelter on the east. Centre right is the terrace of cottages at Lower Works. The railway line to Pontesbury passes a little to the left of Lower Works terrace. In the foreground are Windsor Cottages, on the edge of Snailbeach Coppice. AUTHOR'S COLLECTION

An old postcard view of part of Snailbeach, taken from the Black Hillock. Shop Lane runs between the big shed and the terrace of cottages to the left, while the SDR mine branch passes around and above all the buildings. The Primitive Methodist chapel is in the mid-distance on the right, with the steep roof. Notable are the irregular gardens and the jumble of different roof profiles, pointing to squatter cottage origins.
COURTESY IRONBRIDGE GORGE MUSEUM TRUST

A postcard view from the west of Snailbeach. The largest building is the Methodist chapel, where the railway passes close by, but the Cornish engine house and the Big Chimney dominate the skyline. The SDR mine branch runs behind the cottages on Resting Hill, while the main line passes this side of the chapel. The lineside hedgerows are now well established and envelop the railway fences. AUTHOR'S COLLECTION

partnership became the Snailbeach Mine Co. Ltd by a private Act of Parliament in 1867. Snailbeach became a hamlet and by the 1870s a village, housing mine workers and their families chiefly in squatter cottages.

In the mid-nineteenth century, the mine company began to suffer with the reduction in the price of lead, brought about by the opening up of reserves overseas. The story of the Snailbeach District Railways (or SDR) really begins then. Snailbeach mine still relied on costly horse-drawn carts to link it with the River Severn, and from around 1850 the national railway network, some miles away at Shrewsbury. Finally, in 1861 the Shrewsbury & Welshpool Railway opened its first line from Shrewsbury to Minsterley, affording significant cost savings.

Mineral railway schemes for the Hope Valley followed the building of the Minsterley line. They were mostly impractical, and the heavy investment they needed would have never been forthcoming. The only scheme likely to succeed in all respects, and especially in making local mines more competitive, was the SDR. Promoted by the Snailbeach Mine Co. Ltd, it was enacted by Parliament in 1873, opened for traffic in 1877 and worked for 82 years.

The SDR was essentially a steam-hauled narrow gauge mineral railway, but also a common carrier which carried goods and parcels for the public. As first conceived, its principal traffic was pig lead in the outward direction, and coal for firing the mine engines and smelting furnaces inwards. Unfortunately the railway was built to serve Snailbeach mine just a few years before a major fall in the price of lead. The mine was then too old and deep to compete strongly with large new discoveries of lead elsewhere in the world. Mining was severely curtailed, and traffic on the railway was much reduced as a result. Early in the twentieth century, the railway became increasingly reliant on companies working the spoil heaps at Snailbeach and quarrying stone close by. From the mid-1920s until it closed the staple traffic was road stone.

Part of the reason for the SDR's long life was that its customers were poorly served by roads. Because of this, the quarries which the railway served in the twentieth century were a captive market, but it could not last forever. It had brief competition from steam road locomotive haulage and an aerial ropeway, and then was worked in conjunction with lorries to exploit the road stone under Callow Hill – the line's last source of traffic. The lorries were the writing on the wall. When this last quarry was modernised, the company survived, but not as an operator of railways.

How the Snailbeach District Railways Company operated for so long in the face of these and various other economic disadvantages is the primary subject of this book. But it also highlights the importance of the lead mining industry as a stimulus to railway building in south west Shropshire, and how the SDR served as effective industrial transport which improved the economics of mining and benefited the community.

Snailbeach mine – the 40 Yard Level with 1 foot 10 inch gauge tramway and rotting tramway tubs in 2004. I. COOPER

CHAPTER TWO
The origins of mineral railways

Railway development in Britain

The practical problems and expense of horse haulage to and from Snailbeach mine were nothing new. Mining companies had suffered high transport costs for centuries. Much of the flow of ideas and new developments which led to the modern flanged wheel railway was borne of the need to carry minerals efficiently. Miners and merchants could be crippled by the cost of using horse haulage on poor or non-existent roads. Purpose made plankways for guiding wagons in mines were quite common in the late Middle Ages. This general idea of guiding the wheels led to the building of surface railways for British coal mines in the early seventeenth century, on which flanged wheels became standard and wooden rails were used. Wooden railways were still being built in the eighteenth century, but at the end of the century cast iron plateways and edge railways began to predominate on new lines. Railways ten or twelve miles long were feasible, and were built to deliver goods at a canal or river wharf or a sea port.[1] Shropshire's first recorded railways served coal mines. One was built in 1605 to carry coal from a pit in Broseley about a mile to a wharf at Calcutts on the river Severn. Another short, neighbouring line was built at an uncertain date, but probably in the late 16th century, and served an adit mine to take coal to the same wharf. They were wooden railways.[2]

On railways generally, propulsion turned eventually from horse and gravity to steam. Steam powered haulage first involved stationary engines and then the locomotive. Slowly the advantages of wrought iron rails over cast iron for the carriage of heavy loads became clear, and

SDR Deposited Plan – sheet showing Snailbeach newly surveyed by Henry Dennis. James Ray Eddy's horse tramway with self-acting incline to the smelt works is shown. PARLIAMENTARY ARCHIVES

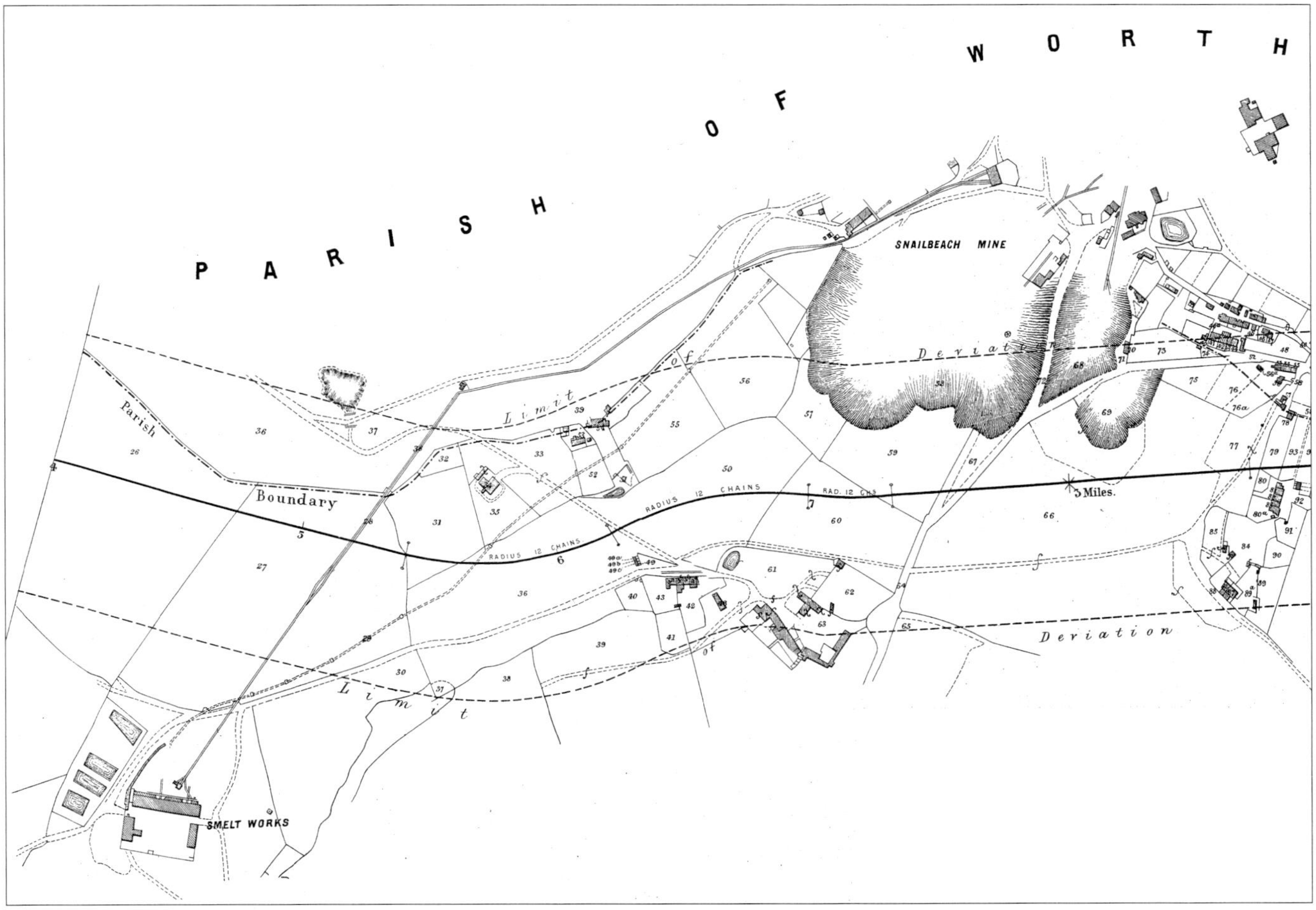

1 foot 10 inch gauge mine tub at Snailbeach about 1959, feeding the spar processing plant run by Joseph Roberts on the White Hillocks. P. G. HINDLEY

Found at Snailbeach, a cast iron double flanged edge rail wheel on a piece of edge rail. D. POYNER

locomotive development forged ahead on an improved edge rail system. The birth of the public railway network in Britain can be said to have begun with the opening of the Stockton and Darlington Railway in 1825 – but it was principally a *colliery* railway, with a passenger service tacked on as an afterthought for those who could afford it. Here, goods were hauled by steam locomotives on wrought iron rails.

The self-contained mine railway (or 'tram') was destined to remain a relatively crude affair, but it was beneficially influenced by these advances. The British mining engineer John Curr introduced his portable track for coal mines in 1787. In effect it was a plateway, but using oak ties (or sleepers) between the plates to tie them to gauge, like a modern railway. The normal plateway of the time used pairs of stone blocks, which did not tie the plates, instead of sleepers. Curr's system included straight and curved plates and turnouts, and the matching sleepers were slotted to accept the plates. The parts were pegged together on assembly, forming short sections of track.[3] In the early 19th century, various slate and stone quarries adopted light and simply made edge rail track systems which used readily available oval or rectangular wrought iron bar as rails. These were fitted and sometimes wedged into slots in wooden sleepers, and short panels of track could be quickly and easily made and moved around. Not so simple was the Thomas Hughes rail, a bar rail having locating lugs formed on it which dropped into drilled holes.

The portable track for agricultural use originated by Robert Neilson was a very

Five different types of rail used at Snailbeach mine. Left to right: cast iron inverted T rail and plate rail, two different wrought iron bridge rails, and a small wrought iron flat bottomed rail. The visible figures on the steel tape measure are inches. D. POYNER

A cast iron tie found at Snailbeach with plate rails assembled to it. It gives a track gauge of about 19 inches over the vertical flanges of the plates. D. POYNER

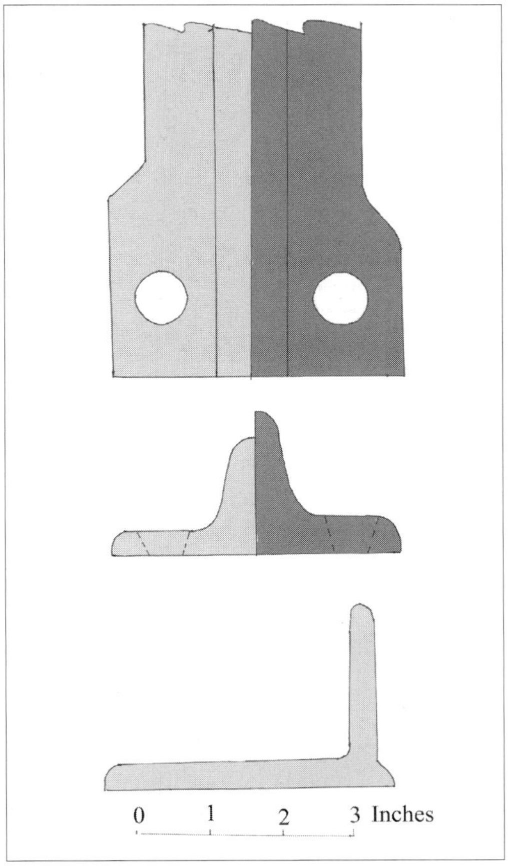

A sketch showing cross sections drawn to scale of (left-hand and right-hand half-sections) the two patterns of inverted T rail and (bottom) the plate rail found at Snailbeach. The top sketch approximates the top views of the inverted T rails. D. POYNER

significant advance. It was first supplied as an extremely popular mass-produced kit in 1848 by iron founder William Crosskill. It had wooden sleepers and rails, the latter faced with angle-iron and usually set at 2ft 6in gauge.[4] The Crosskill product directly inspired Paul Decauville of Petit-

An inverted T rail joint on the 40 Yard Level (crosscut to Chapel Shaft). Shows two rails butted up to each other and nailed to the sleeper which has rotted away. The alignment is not especially accurate, but this made no difference when pushing a tub over it. There are clear perpendicular sleeper depressions across the level in line with every rail joint, and feeling under joints it is still possible to find timber remains. I. COOPER

Bourg, France, who introduced the first all-steel narrow gauge portable track in 1876.[5] This was followed by similar British and continental versions, usually of 60 centimetre (cm) gauge (Imperial equivalent 1ft 11⅝in), and this type of track was commonplace in mining, quarrying and other uses for more than fifty years. Portable track suited the temporary nature of mine railways, which needed to cope with changes to exploration plans underground, reorganisation of surface work and the expansion of spoil heaps. It only took a few men a matter of hours to add some track or change the layout. The mine tub stayed fairly crude and rugged, but small enough to push and to negotiate the tunnels. Where a pony (or later a narrow gauge locomotive) could be used, several tubs could be hauled together.[6]

At Snailbeach

The earliest recorded use of railways at Snailbeach is in the early 1830s, when a gentleman diarist noted that the mine had 'several rail roads', as railways were commonly referred to then.[7] A little is known which suggests the type of railway and its construction materials used at Snailbeach then. Types of rail used at various times for mine tramways have

Scale drawing of inverted T rail measured on the 40 Yard Level (crosscut to Chapel Shaft). I. COOPER

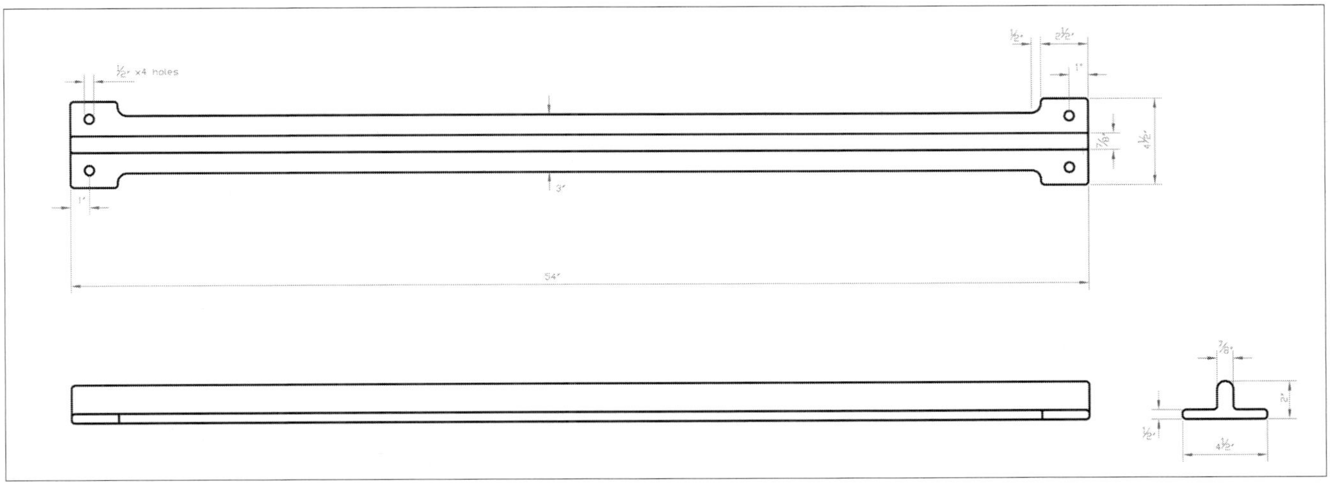

The entrance to Paraffin Level (near Black Tom mine), with track measured as 1 foot 10 inch gauge and two feet between rail centres.

been recorded when found either about the mine or buried in the spoil heaps as odd pieces or even surviving track. They include cast iron plate rail and inverted T rail; and wrought iron bridge rail and flat bottomed rail of various sections and weights.[8]

Plate railways remained popular well into the 19th century, but cast iron inverted T rail was a dead-end innovation in the development of edge railways, which was quickly overtaken by other edge rail advances using wrought iron. It is known mostly for its first use on certain canal feeder railways built about 1800 on the Shropshire-Montgomeryshire border. No firm origin for this system has been pinpointed, but it seems most probable that a quarry company, when building its line to either the Montgomeryshire Canal or the Carreghofa branch of the Ellesmere Canal, set a precedent for neighbouring lines when it bought inverted T rail from one specialist iron founder in the area, possibly William Hazledine of Shrewsbury. It is also possible that the canal engineers William Jessop and Thomas Telford had some involvement. In their original form, inverted T rails had fish-bellied bottom flanges which improved their strength in bending under the weight of loaded wagons. The iron founder who first made inverted T rail might have originated the idea as making better use of the iron in a cast bar rail. The tubs will have been fitted with double flanged 'pulley-style' wheels running loose on fixed axles.[9]

The findings of plate rail and inverted T rail at Snailbeach mine are significant. Nothing is known about precisely when these cast iron railways were introduced to Snailbeach, or their sources, but there are clues. One broken piece each of plate rail and inverted T rail are stored in the locomotive shed, so no-one can be certain when they were put there and whether or not they originated on the mine. But two people remember that a number of plate rails were used as edgings on a flight of steps at the mine before conservation work began, and that other pieces of plate rail were found around Upper Works.

Broken pieces of inverted T rail have been found in the recently excavated tunnel under the road from the dressing floors to the ore house. Yet another broken piece is still in use as the fireplace lintel of a Snailbeach cottage shown on the Tithe map of 1847-1848.[10] The earlier Tithe apportionment (a property list linked to the Tithe map and listing this cottage) indicates that inverted T rail was used at Snailbeach before 1844.[11] A complete spare inverted T rail found against the wall in the 40 Yard Level (crosscut to Chapel Shaft) is 4 feet 6 inches long and has parallel bottom flanges rather than fish-bellied. In this crosscut is a 22½ to 23 in gauge tramway 50 yards or more long, of unknown age, unconfirmed as being laid completely in inverted T rail. But one joint was investigated and photographed which is unquestionably made of this rail.[12] All the inverted T rails mentioned are made from two patterns, evident from measurements which have been recorded.[13] Enough examples have been found in a historical context to indicate that these types of railway were put to use at Snailbeach mine. Either rail type could have been in use well before the documented 'rail road' sighting of the early 1830s, and probably was, but proof has not been found.

The first pattern of inverted T rail has a fish-bellied base and a short vertical flange (the running rail). The second is not fish-bellied but has a taller flange. Three patterns of plate rail have been found. The first is about $4^{1}/_{8}$ inches wide and 60 or more inches long. The other two are $4^{1}/_{4}$ inches wide and 52 inches long. They all have slightly fish-bellied upright flanges – the centres are about $^{3}/_{16}$ inch higher than the ends. An iron tie (sleeper) has also been found which is a good fit to the wider plate rails, giving a gauge across the flanges of about 19 inches. There is also a cast iron twin flanged edge rail wheel.[14]

The earliest dated graphic indication that the Snailbeach company might have had a tramway (although not at Snailbeach mine) is found on the Ordnance Survey's two inch to one mile draft map (O.S.D.) of Shropshire c.1816. Between 1792 and at least 1797, the Snailbeach company had a small colliery just west of Pontesbury, south of the site which would later become Boycott colliery. The O.S.D. map shows the location of the mine, while a tramway from the mine to the Minsterley-Shrewsbury turnpike road is also mapped. In possibly 1813 the Bog mine company leased this land, and it is possible that this company built the tramway, which was lifted by 1820, rather than the Snailbeach company. The land was still owned by the Lovett family in 1825. So unluckily the known facts do not place the tramway at the colliery when the Snailbeach company was known to be operating it.

A map in the popular monograph *The Shropshire Lead Mines* by F. Brook and M. Allbutt shows a tramway from the Snailbeach company's Pontesford coal pits to the road opposite the smelt works in about 1860. No documentary evidence has yet been found to support this idea, and even a map of the colliery drawn in the 1850s does not show a tramway. But there is a well made embankment across the field which might have carried one. It could equally well have been built to make a good road for the company's carts, to take coal to the smelt works and to Snailbeach.

The complete spare inverted T rail found on the 40 Yard Level has parallel bottom flanges rather than fish bellied, as shown in the drawing by Ian Cooper. The other broken samples photographed also have parallel flanges. This suggests they were made by an unidentified foundry which did not understand the importance of the stronger pattern used on the earlier Welsh border tramways mentioned above. One pattern is similar but not identical to rails found on the site of the Morda Tramway, which was little more than a mile from these earlier tramways.[15] However supporting chairs for such rails have been found on the latter site, and this was one way of solving the problem of weakness. At Snailbeach, the same job might have been

A postcard view of Snailbeach mine looking west, with mine tramways in the foreground. The rope worked railway incline serving the pumping and winding engines is in the left background. COURTESY E. ROWSON

done by jamming pieces of stone under the rails to take up the slack.

When introducing the inverted T rail system, the same forms and sizes of wagon body and axle used on plate rail might still have been used, simply by changing the wheels to suit the track, and setting the T rail track gauge to suit the modified wagons. Equally, the various kinds of 'edge rail' on the mine – inverted T rail, bridge rail and flat bottomed rail – might easily have been used together for wagons with modern single flanged wheels to be run on them at pushing speeds. The top and inside surfaces of the non-matching rail heads needed to be aligned, which took skill but was possible. The same was done on the Snailbeach District Railway when joining steel rail to the original wrought iron rail, and this worked well with slow running steam hauled trains. While still in working order, both plateway and inverted T railway equipment might easily have continued in use at Snailbeach mine or the company's collieries, despite their being overtaken by railway developments using wrought iron elsewhere. Independent lines of both types might also have been in use on the works at the same time.

Wrought iron bridge rail became available from about the 1820s, and became increasingly popular as rolling techniques improved, thus making it available in longer more convenient lengths. It could be used to make short lengths of portable track by fastening them to timber sleepers using dog spikes. A sample portable length was found at Snailbeach while clearing the white hillocks.

Wrought iron flat bottomed rail was in use by the 1850s. Since that time, trams of 1 foot 10 inches gauge (two feet between rail centres) and of 2 feet 4 inches gauge (2 feet 6 inches between rail centres) have been used. Some 1ft 10in gauge track survives on the surface and in the 40 Yard Level, while some of 1ft 9in gauge (probably laid that way by mistake for 1ft 10in) is embedded in the road outside Perkin's or Roberts level.[16] 2ft 4in gauge appears to have been one of a range of gauges used by British industrial equipment suppliers (see Chapter Three). Quantities of any of the types of rail found might have been offered to the company either new or second hand by a foundry, or by a mining or colliery company which was liquidating.

When the 60 cm gauge portable track and equipment from Decauville and other British and continental manufacturers had become well established, factory made tram tubs of this gauge became more cheaply and easily available. These would have run easily on two feet gauge track, even when roughly laid. This is the likely reason why newer mine trams installed at Snailbeach, say following the Great War, were sometimes 2ft gauge, and occasional modern, factory made wagons have been identified. The 2ft 4in gauge tubs and all kinds of rails survived as long as they were serviceable and still working on a paying section of the mine, either below ground or on the surface. For instance, George's Shaft was used by the Halvans Co. from about the Great War to raise barytes from above the 112 Yard Level. A short piece of 2ft 4in gauge track was installed in the single winding cage, which brought tram tubs to the surface and allowed them to be pushed out onto a short surface tramway.[17]

A small tramway system was built at Snailbeach in about 1863 by James Ray Eddy. This was a halfway house, of sorts, between a mine tram and a fully fledged narrow gauge railway. James was the son of Stephen Eddy, a highly competent mine manager who was for many years mineral agent at the Duke of Devonshire's Grassington mine in north Yorkshire. James was also well qualified, having trained under his father, completed an apprenticeship to a civil engineer and then managed mineral works alone. In 1857, Eddy and Son took a joint contract to manage the works of the Snailbeach Mine company. James took full charge of the company's works on his father's death in 1861.[18] Eddy's system at Snailbeach was a 3 feet gauge railway using wrought iron flat bottomed rail which linked the mine to the smelt works about half a mile away. It carried the concentrates, or dressed ore, from the ore house to the works, with no return load other than empty tubs. Where the gradients were favourable, the route was horse worked. The remainder was a three-rail self-acting rope worked incline, with a common centre rail and a passing place.[19] Although the route mileage was short, Eddy was cost conscious. Horse haulage was expensive, and their replacement with an un-powered winding drum house was a relatively small investment which would minimise the cost of carriage.

Ordnance Survey maps and photographs show extensive surface tramways at Snailbeach in the late 19th century.[20] In his reports of the time, the mine engineer Henry Dennis still refers

Snailbeach mine in 2004, the 40 Yard Level (crosscut to Chapel Shaft) with 1 foot 10 inch gauge tramway, thought to be laid completely in inverted T rail. Standing against the tunnel wall in the background is the spare inverted T rail measured and drawn by the photographer. I. COOPER

to 'rail roads' in use underground.[21] The Snailbeach District Railways (SDR) had been built by then, and branches were laid to serve the mine and the smelt works. These lines were laid on land leased to the mining company, and were private owner sidings rather than mine tramways. In the early years, the main siding at the mine was directly connected to a 2ft 4in gauge tramway on one of the dressing floors, and this same tramway crossed the main siding on the level on its way to the spoil heaps.

It is debatable whether Eddy would have considered building a railway to the standard gauge railhead like the SDR which followed. He might have had the expertise required, but capital costs were a personal concern to Eddy because of the nature of his contract with the mine company. He received commission on profits, and a large, risky capital project like a three mile railway would have been unlikely to benefit him. Capital expenditure meant reduced profits.[22] Also to be considered was the need for a costly Act of Parliament. His small smelt works railway was contained within the land leased

The drill smiths' shop with loaded 2 feet 4 inch gauge tram. The track is laid in wrought iron flat bottomed rail.
COURTESY E. ROWSON

from the Marquis of Bath.[23] Land acquisition was unnecessary and it was for private use, so it did not need an Act. Besides, James Ray Eddy made large cost savings with the opening of the Minsterley branch railway in 1861 (see Chapter Four). The carriage of lead from the Snailbeach Mine company's old smelt works at Pontesbury to Shrewsbury was reduced from about 6s to 3s 6d or 4s per ton by using the branch railway. The cost of haulage for all freight per ton of lead produced was reduced by around 3s per ton.[24] Conversely Henry Dennis, who succeeded Eddy as mine manager and engineered the SDR, was paid a fixed salary.[25] He had past experience of Acts of Parliament for railways, so there was no obstacle to his considering the idea.[26]

It is also worth exploring why the mine company did not build a railway from the mine to the River Severn at Shrewsbury, the nearest river port, at a much earlier time. Before 1861, haulage between Snailbeach and Shrewsbury, the mine's primary outlet for lead, was an easy task compared with transport out of other lead fields such as the Pennines. The Shrewsbury to Bishop's Castle road via Minsterley (now the A488) was a turnpike, which helped.[27] So although carting was expensive and costs were increasing, it cost less here than in other parts of the country. There was little pressure to take drastic measures to reduce transport costs, such as building a railway. This was especially so while the price of lead remained high. This position seems to have applied to lead producers all over Britain. It appears that few if any railways had been built specifically for the lead industry before the 1840s. The Cromford and High Peak Railway carried a little lead ore at low cost for remote Derbyshire lead mines, but was built principally for coal, limestone and general goods traffic. The Redruth and Chasewater Railway of 1825 carried copper ore and coal. Built by John Taylor in his quest for maximum economy using scientific methods, it showed what railways could do for the metallic mining industry.[28]

Yet even in Stephen and James Ray Eddy's time, the lead trade was fast becoming global. From 1854 to 1877, imports of lead into Britain climbed almost ceaselessly. Over the same period the price of lead followed a general downward trend, with a recovery for only a few years in the early 1870s. In 1877, Warrington W. Smyth, one of HM Mines Inspectors, summed up the position:

> The present period is one of great change, anxiety, and uncertainty in many branches of metallic mining in this country ... within the last two decades the increased activity of Continental and of remote colonial enterprise has brought rivals into the field, who, in some instances, have succeeded in coping victoriously with our poorer and deeper mines ... Turning to the lead mines, ... the British districts noted for this metal ... keep us well in the van of the world, notwithstanding the vast quantities which from time to time have been thrown on the market by the Western States of America, by Spain, and by the island of Sardinia.[29]

The state of the industry cannot have been lost on Henry Dennis. Snailbeach mine was still rich. It had not yet been closed down by overseas competition, but it was being sunk deeper and worked at increasing cost. There was a clear case for adopting any new practice which would significantly reduce operating costs, keep Snailbeach competitive and prolong its life. One possibility was to replace carting with rail transport between Snailbeach and Pontesbury. Surveying a route for a railway was high on Dennis's agenda when the Snailbeach Mine Company Limited employed him.

CHAPTER THREE

The railways and tramways of Henry Dennis

Henry Dennis became mine manager for the Snailbeach Mine Co. Ltd in 1871.[1] He was already an experienced railway constructor by that time. As a young man he qualified as a surveyor and worked for some time in the engineering office of the Cornwall Railway. Work here included surveying on the line. He was also involved in developments at Padstow harbour, taking soundings needed for building a breakwater. A spell in Spain building mineral railways, the dates of which are uncertain, introduced him to the mining sector of civil engineering which later became his principal concern.[2] In the mid-1850s, he built a 3ft 0in gauge tramway in Denbighshire for the mining engineers John Taylor & Sons, between the Llangollen slate quarries at Oernant and a slate mill beside the Llangollen branch of the Shropshire Union Canal.[3] It had upper and lower horse worked sections linked by an inclined plane.[4] The line was completed in 1857, and featured local slate slab in its construction. This was used both in the permanent way instead of timber sleepers and in some structures such as bridges.[5]

Dennis was also engaged to make surveys for various turnpike road and railway schemes to link the Ceiriog valley in Denbighshire, west of Chirk on the Welsh border, to the outside world, notably Ellesmere. His standard gauge Ellesmere and Glyn Valley Railway scheme received the assent of Parliament but was never built. He then surveyed and built the Glyn Valley Tramway (GVT) between Glyn Ceiriog and Chirk, which opened for traffic in 1873. This was a horse drawn 2ft 4in gauge roadside line. It had some steep gradients which made haulage costs excessive, but the real cause of this was the short sighted restriction to horse working. This, and the failure to connect the line to the Great Western Railway at Chirk, made it a rather unsatisfactory line. Neither of these shortcomings were Dennis's fault, but a result of financial restrictions placed on him by the GVTCo.

Also about this time, Dennis engineered and built the horse drawn Wrexham District

November 2009. The bridge carrying the Maes-yr-ychain Lane over the incline on Henry Dennis's Oernant Tramway, looking from the south. The abutments are built from local stone rubble in mortar, and carry heavy slate lintels. The span is only about six feet, to accommodate the tramway.

Viewed from the south in 2009, Dennis's Oernant Tramway viaduct on the approach to the Pentre'er Felin slab and slate works near Llangollen. This is a rubble stone revetted embankment or causeway about 130 yards long. It is more obvious from a distance that the track is raised above the stone work by another graded embankment – it can be seen here. The main arch, second from the left and said to be dated 1857, spans the River Eglwyseg, while there is a flood arch on either side. Dressed stone is used in the arches and their buttresses. There are seven large raking buttresses, on the south side only, which are probably a later addition. The date is quite late for a structure of this kind – more of the early railway era than the modern. Henry Dennis's railway experience spanned both.

Tramways.[6] It was authorised by Act of Parliament in 1873.[7] Unlike the GVT, it was a genuine urban tramway. It was built to 3ft 0in gauge, and opening in 1876 was operated initially with open sided single deck passenger cars having weather curtains. In 1896 the line was taken over by Wrexham Tramways Ltd, then in 1900 by the National Electric Construction Co. Ltd followed quickly by British Electric Traction. The latter group incorporated a new company, Wrexham & District Electric Tramways Ltd, in 1901, which rebuilt it as a 3ft 6in gauge electric line. The company introduced its first motor bus services in 1913, changing its name to Wrexham & District Transport Co. Ltd in 1914 and finally closing the tramway system in 1927.[8]

Railway building was and continued to be a usual part of the mining and quarry engineer's job. Part of Dennis's operation to sink and equip Legacy colliery near Rhosllannerchrugog, Ruabon, included the building of a tramway to carry coal to the main road at Pentrebychan.[9]

Although the GVT perhaps reflected badly on Dennis, his earlier and later experience shows that he was a good railway engineer. Had the GVTCo not restricted cost, it would have been a far better line. The upper section of his Oernant tramway was built with minimum gradients, by

The lifting bridge taking the Oernant Tramway over the Llangollen branch of the Shropshire Union Canal in the 1890s. The tramway from the right has just crossed the road from the Eglwyseg River. To the left is a short siding beside the canal. Across the canal the line enters the Pentre'er Felin slab and slate works.
COURTESY LLANGOLLEN MUSEUM

The New Inn, which was the Glyn Ceiriog terminus of Henry Dennis's horse drawn Glyn Valley Tramway. COURTESY J. MILNER

following the contours as closely as possible between the incline and the quarries. This was best practice and minimised the cost of haulage. He was already surveying the Snailbeach line in 1872 (while the GVT was being built), and it was an eminently workable railway for steam power, with manageable gradients and curves. For the SDR, Dennis even applied to the Board of Trade for permission to reduce the radii of some curves, as he had found that this would enable the gradient to be eased. The reduction in radius was contrary to the Act of Parliament and permission was refused.[10] He was well aware of what made an efficient railway.

The track gauge of 2ft 4in used for the SDR seems rather an oddity in itself. The Snailbeach District Railways Act specified that the gauge be no less than 2ft 4in with no particular upper limit, but that the land taken must be sufficient to accommodate a standard gauge railway, and that all bridges constructed must also be wide enough for such a line.[11] The fact that Dennis adopted 2ft 4in for both the Glyn Valley and Snailbeach lines has been the cause of some speculation. It should be noted that the gauge of the GVT in its horse tramway days is usually recorded as 2ft 4¼in, whereas the contract to build the line specified 2ft 4in. Why was this gauge used for the GVT in the first place, and why did Dennis use it again? The Glyn Valley Tramway Act itself was quite flexible, specifying that the gauge be no less than two feet and no more than 4 ft 8½ in. (For more detailed discussion of GVT track gauge, see *Rails to Glyn Ceiriog – The History of the Glyn Valley Tramway* Part 1, by J. Milner and B. Williams.) One suggestion is that since these various railways were in or near Shropshire, they merely continued the tradition of a 'Shropshire gauge' of around 2ft 4in or larger which became popular for mine tramways in the Severn Valley coalfield.[12] What this fails to address is that the 2ft 4in track gauge was used countrywide. It seems

A scene on the Wrexham District Tramways system, which was engineered by Henry Dennis. The tram car, having full sides and an upper deck, is not an original one.
COURTESY J. MILNER

more likely to have been a gauge which was promoted for a time by some industrial railway equipment manufacturers. Track gauges at four inch intervals – 1ft 8in, 2 feet, 2ft 4in, 2ft 8in and 3 feet – were available, as is evident from records of industrial railways with an early history. Apart from this consideration, clearly Dennis had decided that this gauge of track could handle the anticipated payloads on the gradients and curves which the SDR would have.

It was mentioned in Chapter Two that some mine tramways at Snailbeach were of 2ft 4in gauge.[13] They existed before the SDR was built, and as well as equipment availability, influenced Dennis to select 2ft 4in gauge for the railway, since he saw a need to join the railway sidings to the mine tramways. The Dennis working plan of the railway at the mine shows where the main siding was to cross and be connected to an existing tramway from a dressing floor to the spoil heap, using a wagon turntable which was sketched in roughly and lightly.[14] This would have been impractical, since a wagon turntable was not suitable for being crossed by a locomotive, and a right-angled crossing and two points forming a slip road were installed instead. This is shown in the photograph of the mine c.1885 by J. Thomas, held in the National Library of Wales. This allowed SDR wagons to be loaded with dressed minerals on the dressing floor, eliminating barrowing or other transshipment, and then to be pushed onto the long siding. Mine tubs loaded with waste from the dressing process crossed the siding on the level to go to the waste heaps.

Suggestions have been made that the track gauge of the SDR was actually 2ft 3¾in, but there seems to be no real evidence for this. No railway company records mention any gauge other than two feet four inches. Neither do the Annual Returns to the Board of Trade. The Act of Parliament specifies that the gauge be not less than two feet four inches. There are only two appearances of this figure 2ft 3¾in that the author knows of from attested primary sources. One is found in the records of locomotive builder W. G. Bagnall, concerning the gauge of a Snailbeach Baldwin locomotive under repair.[15] The second is a mention in a letter to the Shropshire County Council from W. H. Austen in 1932.[16] What the figure precisely refers to is not fully established. One suggestion is that it indicates the spacing dimensions used by Bagnall when specially re-setting the tyres of the Baldwin locomotives, which were originally built to run on 60cm gauge track (see Chapter Thirteen). As far as track gauge is concerned, the author measured the track at Snailbeach mine in the 1980s, and found the gauge to be anything from 2ft 4in to 2ft 5in, with the latter occurring where the gauge had spread because of rotten sleepers. The gauge of the line was 2ft 4in, legally and in practice, although the provisions for the Baldwin locomotives which made them useable on the railway were a special case.

Although not by Henry Dennis, another railway of 2ft 4in gauge was built not far from Chirk in about 1874. This existed briefly and was built for Ifton-Rhyn Collieries Ltd. The engineer of this line selected this gauge to match the GVT. He planned to connect his line with Dennis's tramway and transship coal to the Great Western Railway.[17] The Ifton-Rhyn line was the source of the first locomotive bought by the SDR (see Chapter Thirteen).

By the time Dennis joined the Snailbeach Mine Co. Ltd, the modern flanged wheel railway had long since come of age. The Railway Mania of the 1840s had passed, and there was an untidy network of railways totalling thousands of route miles across Britain. Steam-hauled narrow gauge railways were also well proven, and locomotive builders were routinely building relatively powerful narrow gauge locomotives. Henry Dennis was able to combine his own experience with that of others to provide Snailbeach and other local metal mines with a simple but effective railway, and help keep them competitive in the world market.

CHAPTER FOUR

Schemes for Railways*

* Title of a rough note from a Snailbeach Mine Co. Ltd Board meeting

Standard gauge railway schemes

In 1848 to 1849, Shrewsbury entered the railway age with the completion of the line from Chester via Wrexham and Ruabon: the Shrewsbury & Chester Railway. The Shrewsbury & Hereford Railway was enacted in 1846 and completed by 1853. In 1848, a branch line of the Shrewsbury & Chester Railway from Gobowen to Oswestry had also been built, and a spate of railway proposals in the Welsh Marches followed, several of these anticipating a route through central Wales from either Oswestry or Shrewsbury. Two key promotions were the Oswestry & Newtown Railway and the Shrewsbury & Welshpool Railway. It was as a part of this drive westward that the south west Shropshire ore field got its rail head at Minsterley.

By 1855 there were two schemes for a railway to Minsterley. The first was the Shrewsbury & Welshpool Railway (S&WR) which was quickly challenged by the Rea Valley Railway (RVR). The proposed route for the latter was from Montgomery via Chirbury and Minsterley to Shrewsbury, joining the Oswestry & Newtown Railway (O&NR – then being considered by Parliament) to the Shrewsbury & Hereford Railway. As outlined in its prospectus, one of the objectives of the scheme was to serve and profit from the south west Shropshire ore field, which was attracting outside investment following the increased price of lead during the Crimean War of 1854 to 1855. It was also expected to open up markets for the west Shrewsbury coal field.[1] But the RVR scheme was sunk when the S&WR was enacted in 1856. It provided a short branch line to serve Minsterley and a more direct westward route to connect with the O&NR. The S&WR line from Shrewsbury via Hanwood to Minsterley was opened for traffic on 14 February 1861, and between Buttington Junction north of Welshpool

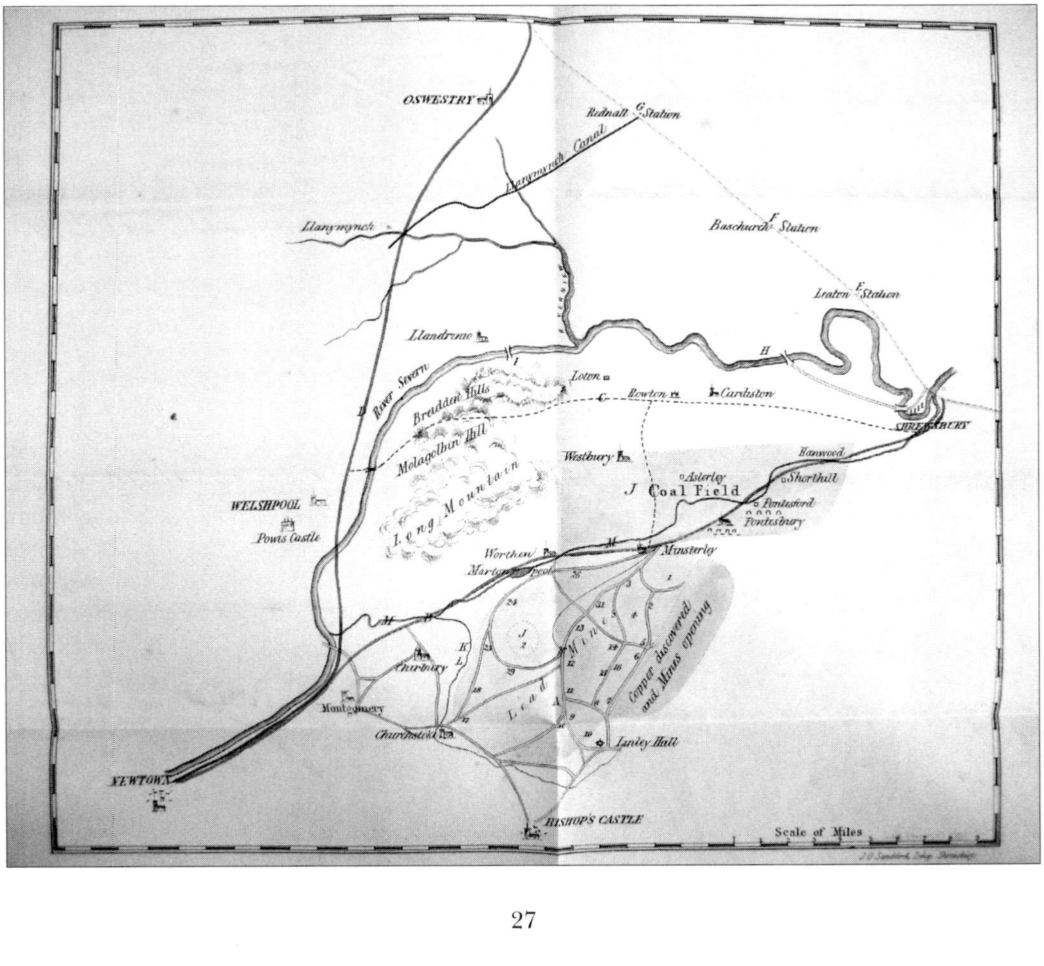

The Rea Valley Railway scheme of the 1850s, which was rejected by Parliament in favour of the Shrewsbury & Welshpool Railway. The reference to 'copper discovery' east of the Stiperstones was optimistic at best. There were no workable veins of copper ore, and any coppery mineral found there is the result of chemical reactions around other minerals. Such mineralogical knowledge was not widespread even at that time. Miners such as the Harrison brothers and John Job of Snailbeach, who tried mining at Westcott, were searching for copper but would have found only barytes in any quantity.
THE NATIONAL ARCHIVES

Lydham Heath station on the Bishop's Castle Railway – the often suggested southern terminus for railways through the Hope Valley and Shelve.

and Hanwood on 27 January 1862. The S&WR was purchased by the London & North Western Railway (LNWR) under an Act of 14 July 1864. It became vested jointly in the Great Western Railway and the LNWR on 5 July 1865. The two latter companies formed a Joint Railways Committee to administer this and many other Welsh border railways.[2]

The official opening of the Minsterley line took place on 13 February 1861. The railway company gave free rides, while an official dinner was held at the Miner's Arms, Minsterley, presided over by William Burr, the lead works proprietor of Shrewsbury. Also attending were John Job and Jonathan Harrison, the two senior local officers of the Snailbeach Mine company.[3] The line was to be mutually beneficial to the S&WR, the Burr brothers and the Snailbeach company – the latter being the two major local players in the lead market. In this new railway age, T. W. & G. Burr (and later George Burr alone) was the best customer for Snailbeach lead.[4]

Pontesbury station looking towards Minsterley. Pontesbury Junction sidings are behind the low hill in the background.

The S&WR also planned a line from Minsterley via Worthen and Marton to Chirbury, partially repeating the O&NR scheme, and from there via Church Stoke, Snead and Lydham to Bishop's Castle. This scheme was not enacted.[5] Instead, the Bishop's Castle Railway (BCR), authorised in 1861, built a line from Craven Arms to Bishop's Castle via a branch from Lydham Heath which opened in 1865. It also had powers to build a line from Lydham to join the Cambrian Railway near Montgomery (the former Oswestry & Newtown line), and a line between Chirbury and Minsterley, covering the S&WR scheme. Funds were always short and rails were never laid past Lydham. The financial difficulties of the BCR worsened and it fell into receivership in 1867. A Bishop's Castle & Montgomery Railway group revived the Lydham to Montgomery scheme alone, but was dismissed by the Parliamentary Committee.[6] The BCR survived in receivership until 1935.

So in 1865, at the conclusion of this flurry of activity, the south west Shropshire ore field was served by one fairly convenient railhead at Minsterley, giving access to the national rail network via Shrewsbury. Lydham Heath was somewhat far to the south, and a railway through the ore field to join the BCR there would have needed huge investment to deal with the hilly terrain. Any such line would probably have been uneconomical, but such schemes did arise later in the guise of mineral railways. Though all the standard gauge proposals had instead chosen a route through easy country to the west, even the single authorised scheme could not fund itself. Meanwhile the mining interests were left to deal with local transport as best they could.

Mineral railway schemes

Plans were being made for railways in the ore field before Henry Dennis arrived. They were all to take advantage of the newly opened Minsterley branch of the Shrewsbury & Welshpool Railway. Although the price of lead was slowly falling in the 1860s as the lead trade globalised, there was confidence in long term prospects locally. Output was maintained at Ovenpipe (later known as Tankerville), Stiperstones, Roman Gravels and Snailbeach mines, with the district usually raising over 4,000 tons of lead ore per year.[7]

This map of the early standard gauge railway schemes shows how they were confined to low lying areas, while the later Snailbeach line penetrated much higher ground where the Shropshire ore field was found.

The earliest mineral railway scheme known was by R. Jasper More of Linley Hall. It is mentioned in a letter dated February 1862 by James Ray Eddy.[8] Nothing else is known of this scheme, but More was involved in others later. The Mores were landlords of Roman Gravels mine. The second known scheme was in 1866, when a prospectus was drawn up for the Stiperstones (Act obtaining) Railway Company Ltd, intended '... *to enable the Promoters to raise the requisite funds for applying to Parliament ... for an Act authorising the construction of a Railway from or near Pontesbury to or in the direction of Bishop's Castle, with Branches and Tramways.*' It was primarily intended to serve the mines. The promoters suggested that up to £100,000 would be needed to build a line. They sought £2,000 in £1 shares to survey and investigate their proposal, but there were no takers.[9] The proposal followed closely the opening of the Bishop's Castle Railway, and coincided with the recent formation of the Stiperstones Company Ltd, which concentrated on developing Pennerley mine. Much money was spent but the mine remained unprofitable. Small wonder that the railway got no further than an idea. In time more railway schemes to join Pontesbury to Lydham via Shelve would follow (see Chapter Five).

In March 1871, Henry Dennis had just begun working as manager of the Snailbeach Mine Co. Ltd.[10] He ran his mining and civil engineering business from offices on High Street, Ruabon, Denbighshire, but under this contract he attended Snailbeach mine regularly, and was always available for advice.[11]

The Dennis office building was built of red and white Ruabon brick with terracotta decoration, all supplied by various of Dennis's industrial concerns. The offices took the telephone number 'Ruabon 1'. The entrance doors were of mahogany, and opened onto the entrance hall which had a mosaic floor incorporating the name DENNIS in the design. The staircase and all other doors were also mahogany. Various styles of tile, also from Dennis's industrial interests, were used to decorate the floors of ground floor rooms and for wall panelling. On the first floor was the board room with a Spanish mahogany table, and a number of drawing and administrative offices.[12]

Samuel Harley Kough, a solicitor of Shrewsbury and Church Stretton, was working then on a proposal to build a 3ft 0in gauge railway into the ore field, and wrote to Dennis in March 1871 in the hope of him becoming involved.[13] This was the South Shropshire Railways scheme. Kough had a special interest, being involved in local mining companies. He was a shareholder and secretary of the Central Snailbeach Mine Co. Ltd. He also held shares in its successor the New Central Snailbeach Mine Co. Ltd of 1869.[14] Both Kough's scheme and Dennis's which followed would have served the New Central company, which opposed that of Dennis. Kough had the financial backing of a civil engineer named John Keane. The major object of the company was described as making a railway from the parish of Pontesbury to a point or points near Pwll-le-gate in the parish of More. It was to pass through or near '... *Pontesford, Pontesbury, Nills Hill, Minsterly [sic], skirting all or some portions of the North Western flanks of the Stiperstones and Shelve mountains, approaching, passing and touching the various mines or mining setts, quarries, and other industries, situate on, in, and under the parish of Pontesbury, and the mineral districts or areas known as the Snailbeach, Stiperstones and Shelve Districts, and other the adjacent neighbourhoods and parts of the said country.*'[15]

This poor copy of a photograph taken at Minsterley is included to show the character of the joint line in the Victorian era. A passenger and a goods train are shown, and the locomotives and rolling stock are of the London & North Western Railway. The Minsterley branch closed for passengers on 5 February 1951 and for goods on 1 May 1967.
SHROPSHIRE ARCHIVES PH/M/1

The *Mining Journal* of 5 August 1871 briefly discussed the topic of railways in the Shropshire lead field, noting that Kough had publicised the South Shropshire Railways in a recent Shrewsbury newspaper. The company was to be incorporated by Parliament with a capital of £60,000 in 12,000 shares of £5 each. The editorial continued:

> ... more than one scheme has been projected; of these, another is the 'Minsterley Mineral Railway', which appears, from the information afforded by a prospectus issued last week, to be intended to form a continuation or extension of the present 4ft 8½in guage from Minsterley. The capital of this latter undertaking is £80,000, in 16,000 shares of £5 each, and its promoters state that, *'from careful computation, the actual present traffic will pay £5 per cent on the capital ...'* At this early stage we do not think it desirable to offer any opinion as to the relative merits of the two undertakings, but we doubt not they will ere long be fully discussed and ample opportunity afforded for deciding which will best satisfy the requirements of the district.[*sic*][16]

The Minsterley Mineral Railway scheme was promoted by a provisional board of directors including R. Jasper More and another major landowner Charles Spencer Lloyd, and engineered by T. C. Townsend, a civil engineer of Shrewsbury. With approximately 11 miles of line, it was to serve most mines in the field, '... *into nearly all of which sidings may be made.*'[17] More and Lloyd both had major lead mine properties which had begun to improve significantly following investment by joint stock companies – notably More's Roman Gravels – which both schemes would have served. The expectations of the various promoters must also have been raised by the Franco-Prussian War of 1870 to 1871. The price of lead rose for a couple of years and there was a short economic boom.[18]

Perhaps Dennis read the *Mining Journal* article. In his manager's report of 31 August 1871, he wrote:

> I ought not probably to conclude my report without referring to an agitation that exists with reference to Railway Communication to the different Mines in the Snailbeach District. Several schemes have been partially launched, and it will require your careful consideration how far your interests and convenience are consulted by them.

Dennis already had his own opinion. In a letter to John Jones on 10 September 1871 he wrote:

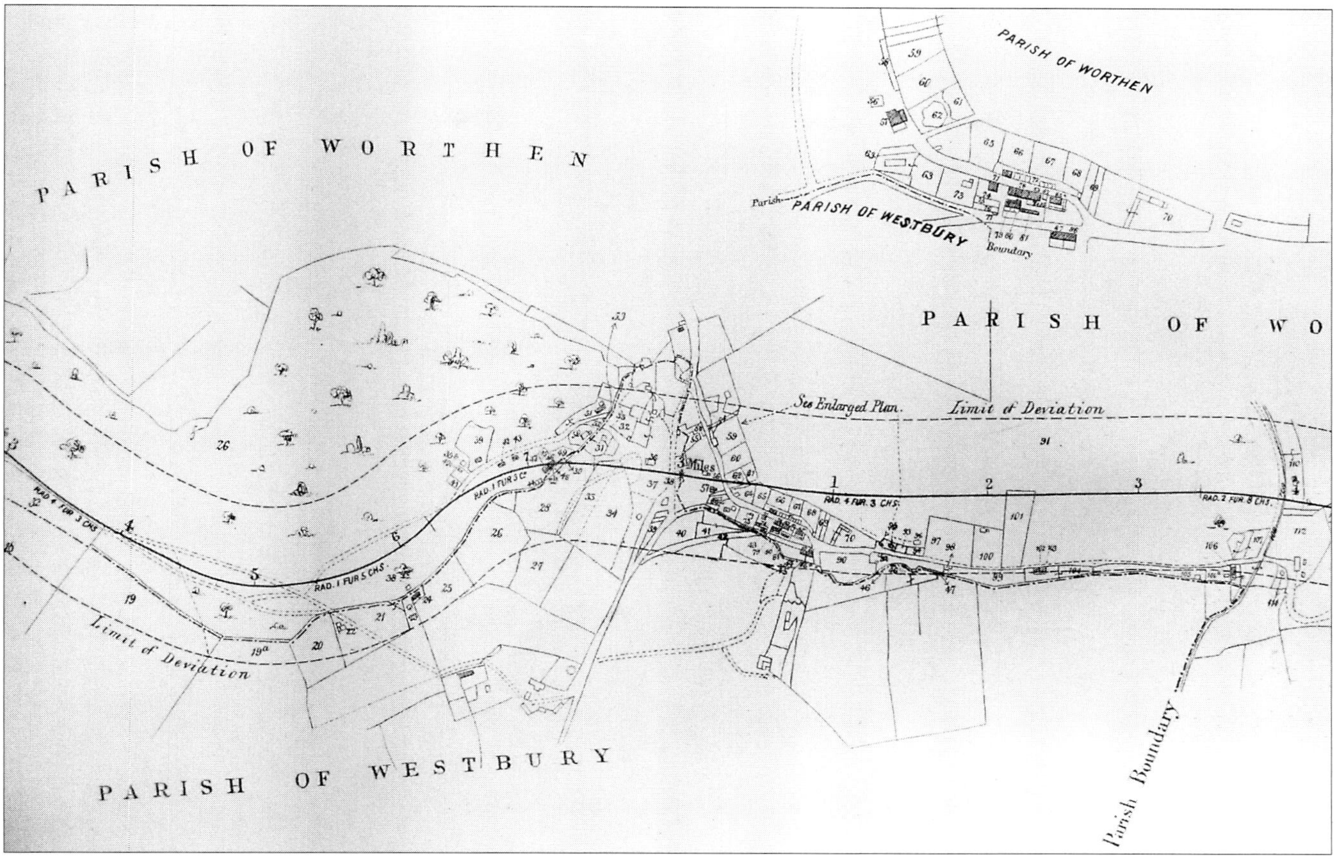

Part of the Deposited Plan for Samuel Harley Kough's first South Shropshire Railways scheme, showing the destructive route through Snailbeach mine. The plan was produced by tracing the Tithe Map c.1848 and does not show accurately Snailbeach mine in the 1870s.
SHROPSHIRE ARCHIVES DP 469

> *Railway*. The *future* development and well doing of these Mines will in a great measure depend on this Railway question – & I must beg of you not to lose sight of it. Mr. Townsend's line is of no use to us whatever – and what is worse if he is allowed to get his Bill – we are prevented from helping ourselves – and to stop them or him from going on with such a mad scheme would be a kindness to them, and to the Mineral Owners of the District. To work Callow Hill we are bound to have a Railway communication with it & Snailbeach, and not to work it would be a grave mistake for I have a very high opinion of the Vein there.

The first indication that Dennis himself had a proposal is a brief mention in the board meeting minutes of 26 September 1871, when '... *Schemes for a Railway were considered* ...' [19] In correspondence Dennis mentioned to Kough the intention to construct a line to Snailbeach, then referred him to John Jones. Kough suggested that the two parties might work together, but pressed on with his own plans, and published shortly after in the *Shropshire News* a notice to obtain an Act of Parliament.

The Joint Railways Committee was also forming an opinion. Patchett, the Minsterley branch station master, investigated the various schemes. In a letter dated 2 August 1871 to the joint superintendents, he favoured Dennis's proposal which would cost much less than the other two. Dennis had given him an approximate cost of £25,000. He also reported that the Snailbeach Mine Co. generated the majority of the present traffic on the branch line from the ore field, and suggested that other mine companies might easily use Dennis's line.

A South Shropshire Railways plan was deposited, according to Parliamentary procedure, for public inspection in Shrewsbury and at Parliament on 1 November 1871. The scheme had an admirable gradient profile, but would only have been achieved at great expense and practical difficulty in construction. Any railway scheme submitted for obtaining an Act of Parliament was divided into numbered sections to make description easy. The route of Railway No.1 of the South Shropshire Railways as far as Snailbeach mine had a similar gradient to Dennis's line as surveyed later of 1 in 37, but it was to be driven on the straightest line possible and slightly further to the east, on the flanks of the Stiperstones. Extensive earthworks would have been necessary for the entire 9¼

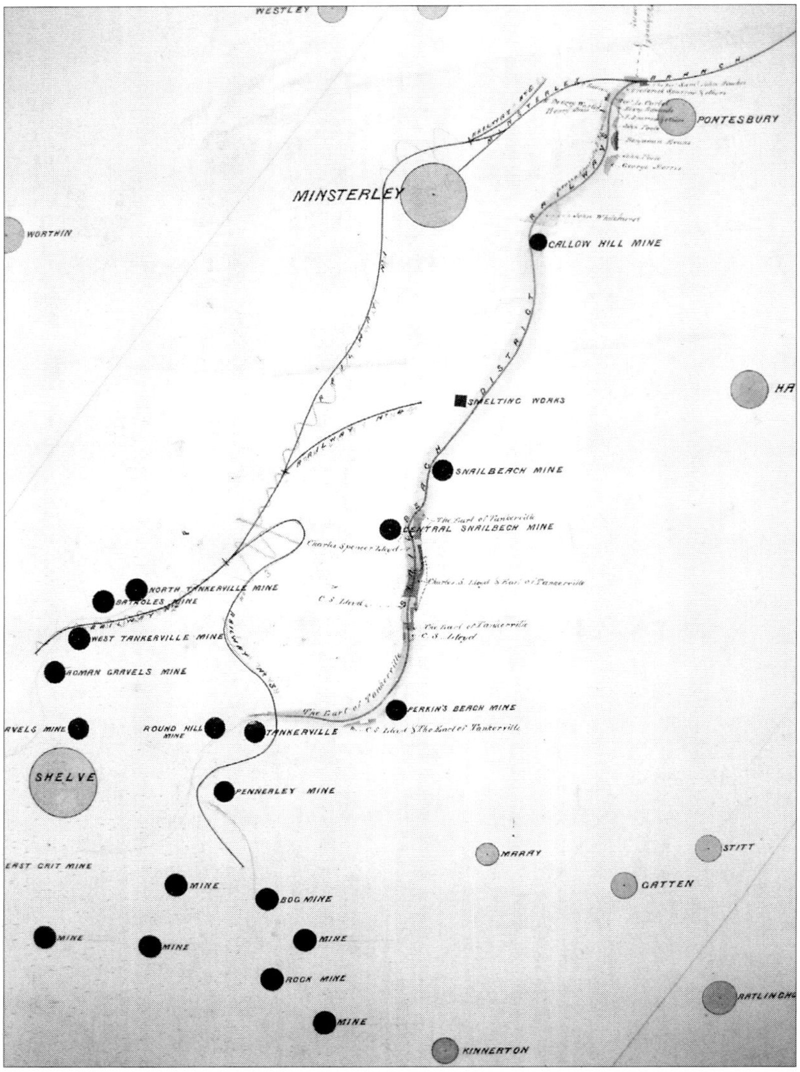

Henry Dennis's railway schemes map, including Samuel Harley Kough's second South Shropshire Railways scheme as well as Dennis's own for the Snailbeach District Railways. This was produced for the mine company and used to discuss railway schemes at board meetings.
The pencil scribblings were probably made when the board meeting suggested to Kough that he make a shorter scheme to join Dennis's at Tankerville mine.
SHROPSHIRE ARCHIVES 4743/47

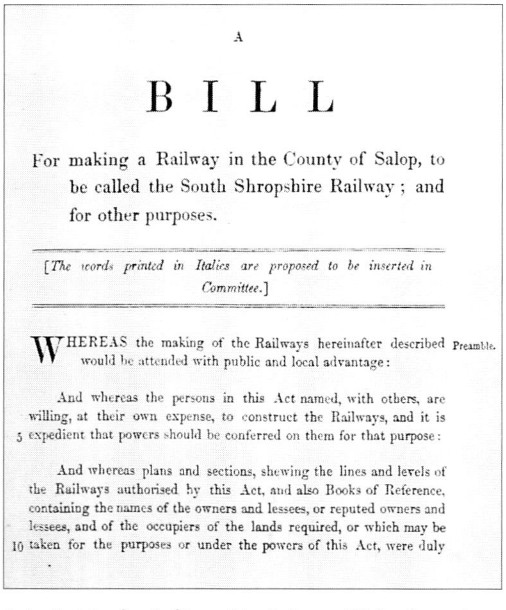

A

BILL

For making a Railway in the County of Salop, to be called the South Shropshire Railway; and for other purposes.

[*The words printed in Italics are proposed to be inserted in Committee.*]

WHEREAS the making of the Railways hereinafter described would be attended with public and local advantage: Preamble.

And whereas the persons in this Act named, with others, are willing, at their own expense, to construct the Railways, and it is expedient that powers should be conferred on them for that purpose:

And whereas plans and sections, shewing the lines and levels of the Railways authorised by this Act, and also Books of Reference, containing the names of the owners and lessees, or reputed owners and lessees, and of the occupiers of the lands required, or which may be taken for the purposes or under the powers of this Act, were duly

Extract of the South Shropshire Railways Bill for Samuel Harley Kough's scheme No.2.

mile length of Railway No.1. Snailbeach mine was to be passed through with supposed convenience in a deep cutting. Many other mines to be served were to be similarly slighted. Road crossings by bridge or level crossing would have been ridiculous. Railway No.2, branching off east of Shelve at around the six mile mark, was rather better: a line with shallow earthworks of about 1¼ miles to serve the Bog mine.[20] Townsend's Minsterley Mineral Railway scheme was perhaps more practical, although Dennis thought it was 'mad'. Attention had been paid to the terrain (although a gradient profile has not been found), but the line passing Snailbeach was only to serve the smelt works. Maximum benefit for the Snailbeach Mine Co. would have only been obtained by this scheme also serving Snailbeach mine, and by passing closer to Callow Hill mine.

Dennis, with the company chairman and the mine company's solicitors Longueville & Co., opposed Kough's scheme by quickly lodging a Petition on Standing Orders in Parliament. Kough wrote again to Jones saying that the scheme had been abandoned, and '... *that if my line afforded proper facilities that was all* [Kough's patron's] *clients wanted.*' Kough also asked to discuss preliminaries '... *in the promotion of this* [new] *undertaking which will be of the ordinary guage.*' [sic] A second South Shropshire Railways plan – standard gauge – was on the way.

By the June 1872 board meeting, Dennis's ideas for a line were becoming firm. He had recently begun surveying a route, and in a draft report dated 8 June, he wrote: '... *if parties in the interest of the Company could be found to initiate a scheme of moderate dimensions and cost, I am satisfied that it would be for the interest of the Company to support it.*'[21] And the Snailbeach line turned out to be just such a railway. The board agreed '... *That it was desirable that a separate Company be formed for promoting it taking powers to the Company to subscribe and that the shares be first offered to parties interested in the Company.*'

The subject was resumed at the proprietors' meeting later in the month. Although we have no record of a discussion of Dennis's scheme, his report to the meeting featured a review of one of the competing schemes.

> The Railway Scheme ... upon examination [*was*] ... very defective, and the petition which was subsequently presented by you against it had the effect of inducing its withdrawal, and fortunately at a very small cost compared with that which would

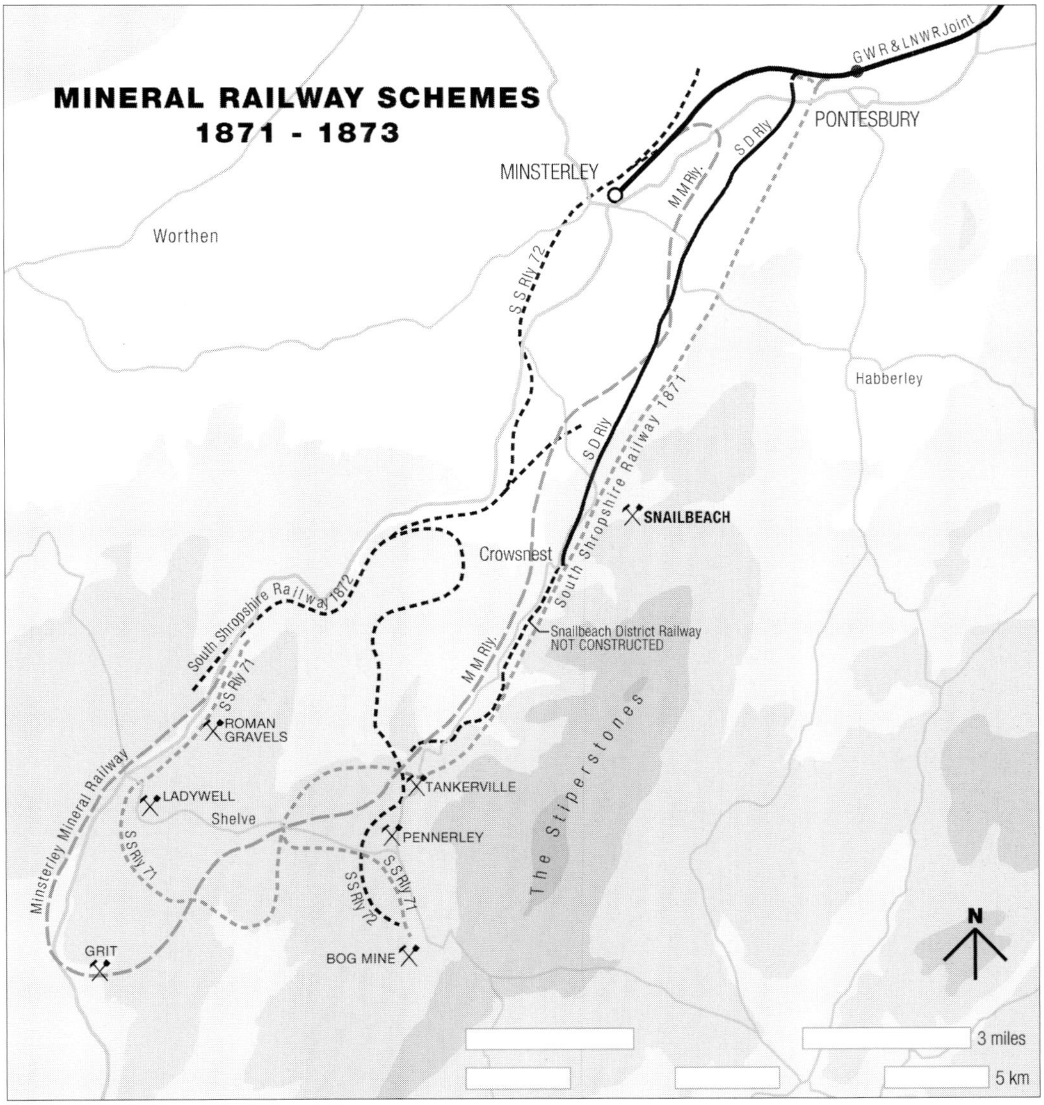

Mineral railway schemes 1871-1873.
Some of the mineral railway schemes mentioned in the text are not indicated by this map. They all would have had similar routes between Shelve and Lydham Heath, and would have served various mines depending on the sponsoring landlord or company. There was also the Boycott Colliery tramway proposal detailed in Appendix Five.

> have been inevitable had it been allowed to advance to a more mature stage. I feel it my duty to forewarn the Company that similar schemes will be promoted in the approaching Session [*of Parliament*] ...

The directors' report to this meeting reinforced Dennis's warning, and reflected his statement of 8 June:

> Having regard to the activity of mining operations in the locality it is only too probable that future Sessions will produce successive crops of similar impracticable schemes, to result either in the locking up of the district or the injurious limitation of your own freedom of action in the transport of material to and from the Mine. Some of your Proprietors are considering the question whether it may not be best to anticipate events by assisting in the construction of a line of suitable pretensions, which they are advised would be remunerative.

Longueville had begun work on the promotion of a Parliamentary Bill for Dennis's railway, and a board meeting in October resolved that the company be recommended to subscribe £2,000 to the capital of the line from its funds. Dennis had by now fully estimated the cost of construction at £20,000, and the Parliamentary Plans were deposited on 31 November 1872. The scheme included an extension beyond Snailbeach to serve the Tankerville mine.[22]

Samuel Harley Kough continued to work with J. H. Tolmé, one of the two engineers of his first scheme. Both men attended the Snailbeach Mine Co. board meeting on 18 October 1872, where they outlined South Shropshire Railways scheme number two, which was to be standard gauge. The plans were deposited in November. The board learnt that the scheme was not intended to serve Snailbeach or Callow Hill mines, although a branch was to be laid to the Smelt Works. Kough was told that they would not support him, but they also suggested that he and his associates might consider building a line to join Dennis's route at Tankerville. The chief mines to be served by Tolmé's plan were Roman Gravels and Bog, and because of their locations relative to Tankerville it was a fair suggestion, in that it would result in one line serving all four major mines with the least route mileage.[23]

The directors reported that same day to the special general meeting about the danger from competing railway schemes, and that

> ... A small Company, having for its primary object the accommodation of your Mines and Works (the largest and most important in the district), has therefore been promoted with the approval of your Directors, all of whom take a large pecuniary interest in it. The principal portion of the necessary Capital has already been subscribed.

The shareholders approved the subscription of £2,000 to the railway, and that the Bill must include power for them to appoint a director to the railway company. The draft Bill was fully approved at two extraordinary general meetings in February 1873. Meanwhile two major landowners on the route, Louisa Hope Edwardes and Lord Bath, had objected to the Bill. In March, his Lordship concluded an agreement with the mine company withdrawing his objection. He would also on request join with the company to oppose any competing scheme.[24] The other schemes looked good for the other major mines, but were of no use to those on his land. Another feature of the agreement was that he elected to allow possession of the land for the railway in exchange for a Chief Rent under the Lands Clauses Consolidation Act. The rent would be charged on the traffic tolls collected by the railway.

The Snailbeach District Railways Bill was making good progress, and Dennis was hopeful that Kough's scheme would not comply with Standing Orders of Parliament. Some time had been spent checking the details of Tolmé's plans during December and January.[25] At the general meeting of 20 June 1873 Dennis confirmed the defeat of Kough's second scheme:

> It was, in fact, full of blunders, and my only regret is, that it should have occasioned so much vexatious trouble and unavoidable expense.

Regardless of the objections raised in Parliamentary Committee and all that trouble and expense, it would again have been an expensive white elephant. Kough had told the board it would cost £6,000 per mile again, but the Engineers' Estimate of Expense showed an average of around £9,000. The total estimated cost was £95,000. This capital was to be covered by shares valued at £72,000 with powers to take mortgages for the remaining £23,000. Such a scheme was doomed to failure.

The Snailbeach District Railways scheme

Henry Dennis was under contract to manage the Snailbeach Mine Co. Ltd's works giving his best professional judgement and advice, and had looked carefully at all aspects of mining at Snailbeach and how to get the maximum return out of the business. Although transport was much improved since the opening of the railway to Minsterley in 1861, the company still used horse drawn carts between the mine and smelt works and Minsterley station, which was difficult and costly. Every mine in the field shared this difficulty.

The mine company directed that the railway be built by a separate company. There were a number of reasons why. Firstly, large amounts of money could not be taken from mining revenue without injuring the shareholders' profits. Secondly, the shares in the mine company were fully paid up, so there was no capital available.[26] Thirdly, it did not own the land for the railway, and an Act of Parliament would be needed to form a public railway company with powers to take land for the line and act as a common carrier. Aside from any other requirement, common carrier status would have been needed for other mine companies' goods to be carried at Parliamentary carriage rates.

The company's prospectus, issued only in December 1875 and for private circulation, outlines what it saw as the salient needs for a railway:

> The object of the promoters is to provide adequate Railway accommodation on a gauge of 2 feet 4 inches, and at a moderate cost of construction and working, for the large and increasing traffic of the Snailbeach Mining District ... [*Until now*] all produce ... traversed the hilly and indifferent roads [*found locally*] ... All [*the mines of the District*] are suffering from the want of easy and economic transport ... [*Apart from mining traffic*] the general production of the district from agriculture and trade, and its wants in lime and fuel, are

> considerable. Passenger traffic will also eventually be an important element, although it has not been taken into present account, because, in the first instance, it is proposed to construct the line for minerals and goods only. ... The tonnage of produce passing to and from the existing Railway system, and the cost of its carriage, with other necessary statistics, have been very carefully ascertained ...[27]

The potential of the railway for profit was looked at closely. The question of building the railway to maintain the profitability of Snailbeach mine (mentioned in Chapter Two) was a separate, underlying issue. The actual savings achieved in mine transport costs can be estimated from historical data. The cost of railway transport between Snailbeach and Pontesbury in 1879 was less than horse haulage to and from Minsterley in 1862 by 1s 5d (or better) per ton of lead produced – about ten percent less. This includes transport of all materials needed for mining, including coal and timber, and in fact includes rail haulage for other mine companies which cannot easily be separated. The shareholders earned a three percent dividend for some years.[28]

Dennis was confident that a railway would give the mine a competitive edge. In 1876 he anticipated, for example, a greater demand for barytes with the opening of the railway. Efficient transport would enable the product to be offered at a more competitive price. In 1878, after the line opened, Dennis announced new customers for barytes paying higher prices. Although the improved transport undoubtedly played its part, general demand for barytes was increasing too.

On 5 August 1873 the Snailbeach District Railways Act became law, and the first general meeting of the Snailbeach District Railways Company was held on 4 November. (The Incorporation by Act of Parliament of the company with this name has in later years given rise to the railway being known in print in the plural: 'Snailbeach District Railways'. It is commonly known in the singular, Snailbeach District Railway, and has been named this way in official papers including the company's 1875 prospectus and on Ordnance Survey maps.) This marked the end of the mine company's control of railway matters for the time being. But as a major shareholder in the SDRCo it was destined never to be out of the picture. The decisive action by the solicitors combined with Dennis's technical knowledge had seen off all opposition, and cleared the way for the building of an effective railway into the district at a reasonable cost. The construction of it resulted in some seriously furrowed brows among the mine company directors along the way.

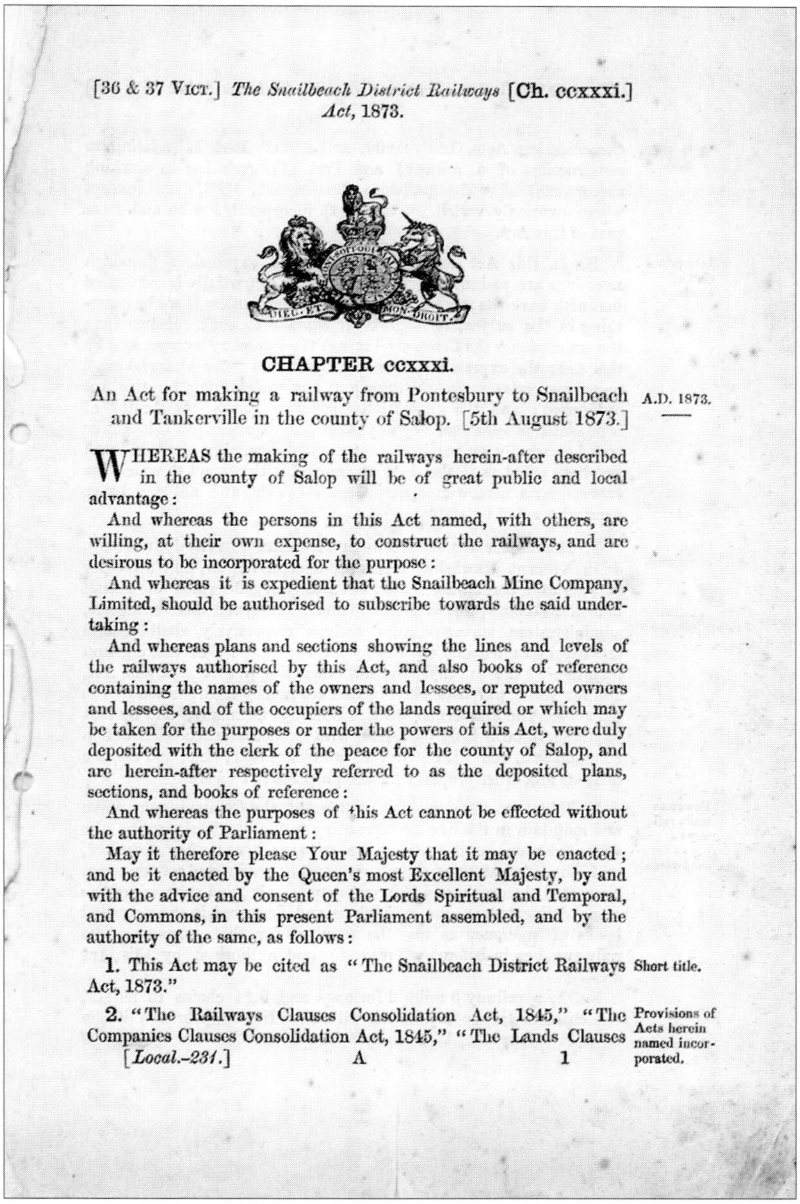

[36 & 37 VICT.] *The Snailbeach District Railways* [Ch. ccxxxi.]
Act, 1873.

CHAPTER ccxxxi.

An Act for making a railway from Pontesbury to Snailbeach and Tankerville in the county of Salop. [5th August 1873.] A.D. 1873.

WHEREAS the making of the railways herein-after described in the county of Salop will be of great public and local advantage:

And whereas the persons in this Act named, with others, are willing, at their own expense, to construct the railways, and are desirous to be incorporated for the purpose:

And whereas it is expedient that the Snailbeach Mine Company, Limited, should be authorised to subscribe towards the said undertaking:

And whereas plans and sections showing the lines and levels of the railways authorised by this Act, and also books of reference containing the names of the owners and lessees, or reputed owners and lessees, and of the occupiers of the lands required or which may be taken for the purposes or under the powers of this Act, were duly deposited with the clerk of the peace for the county of Salop, and are herein-after respectively referred to as the deposited plans, sections, and books of reference:

And whereas the purposes of this Act cannot be effected without the authority of Parliament:

May it therefore please Your Majesty that it may be enacted; and be it enacted by the Queen's most Excellent Majesty, by and with the advice and consent of the Lords Spiritual and Temporal, and Commons, in this present Parliament assembled, and by the authority of the same, as follows:

1. This Act may be cited as "The Snailbeach District Railways Act, 1873." Short title.

2. "The Railways Clauses Consolidation Act, 1845," "The Companies Clauses Consolidation Act, 1845," "The Lands Clauses Provisions of Acts herein named incorporated.

[*Local.-231.*] A 1

The Snailbeach District Railways Act page 1 – the company's original copy.
COURTESY M. J. IGGULDEN

Looking north at the Callow Lane over bridge. All three of the over bridges attracted criticism from the Condover District Highways Board. This bridge must have been especially difficult as the lane climbs very steeply, and the uphill approach can only have been made worse here, where the railway crossed at a level little different from the existing road surface.
H. F. G. DALSTON

CHAPTER FIVE
Construction of the line

With the passing of the Snailbeach District Railways Act on 5 August 1873, the company existed in law. The Act authorised the company to raise construction capital by 2,000 ordinary shares of £10 each, totalling £20,000. Borrowing powers were also empowered to a maximum amount of £6,600. The railways to be constructed were firstly Railway No.1 of length 3 miles 2 furlongs and 5.44 chains, from a junction on the Shrewsbury & Welshpool Railway at a point near Pontesbury station to a point '... *near the highway leading from the Bog and Pennerley Lead Mines by Snailbeach to Minsterley* ...' This was a few yards south of the planned weighbridges at Snailbeach wharf. Secondly, to join at this point, was Railway No.2 of length 1 mile 7 furlongs and 3.90 chains to a terminus in '... *a field belonging or reputed to belong to the Earl of Tankerville, and in the occupation of Richard Thomas* ...' which was a short carting distance from Tankerville mine.

At the first general meeting of the shareholders on 4 November, the main business was to confirm the board of directors. The five named in the Act were Thomas Heaton Lovett, John Henniker Lovett, John Jones, John Vincent Hawksley Williams and Joseph Warter, while Joseph Whitehouse represented the Snailbeach Mine Co. Ltd. They were all directors or senior shareholders of the mine company. Warter's place was taken by another of his family. The secretary's salary was set at £50, to be raised to £100 '... *if the Co. gets into full work.*' The directors elected Thomas Heaton Lovett the chairman, John Jones secretary, and Croxon & Co. of Oswestry the bankers. Jones was a partner of the Oswestry solicitors Longueville & Co., and their office in Upper Brook Street, Oswestry, became the registered office of the SDRCo. The first call of £2 per share was to be made early in the new year. The board was also told of the negotiations between R. Jasper More and Longueville concerning a possible extension to meet the Bishop's Castle Railway.[1] There had been a meeting between the Lovetts and More in

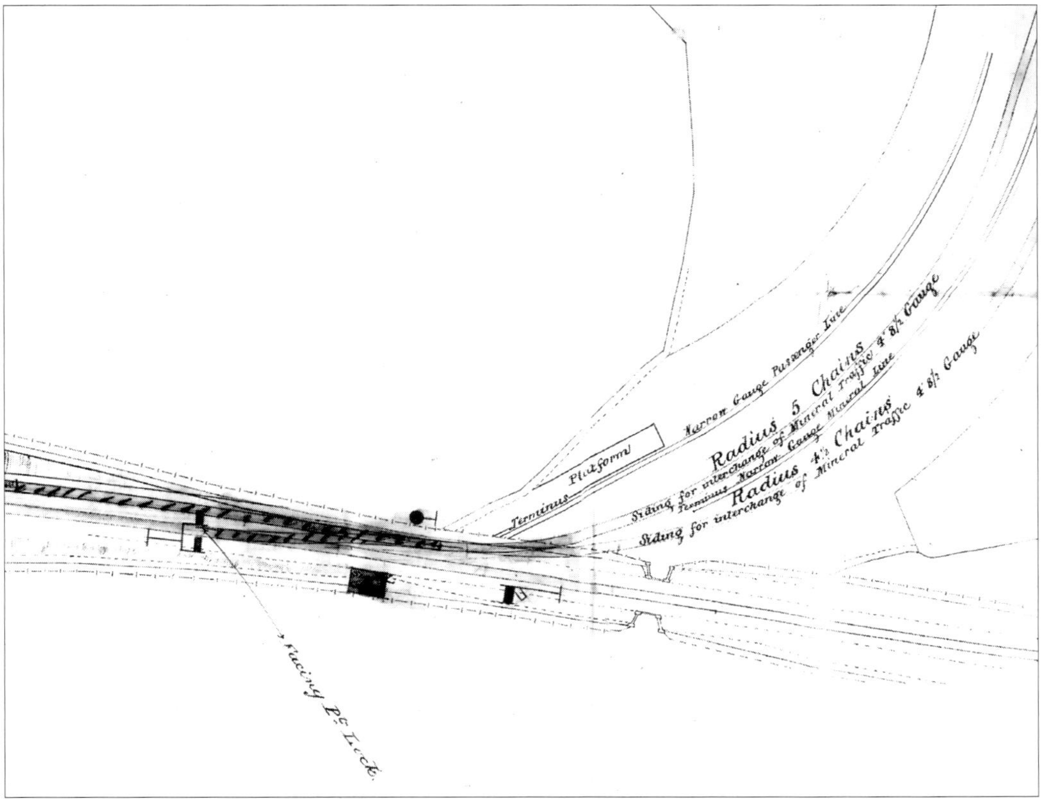

Notable on Dennis's plan for Pontesbury Junction are the 'narrow gauge passenger line' with a platform, and the footpath along the cutting to Pontesbury station. He also felt that only one standard gauge crossover was necessary – the Joint Railways disagreed. Another point was added for a siding to the SDR warehouse.
THE NATIONAL ARCHIVES MT6/180/6

October, and the SDRCo seriously considered the scheme.[2] A civil engineer, George Wilson, was hired by More and he made a number of surveys. The Midland Railway Carriage & Wagon Co., a major creditor of the Bishop's Castle Railway which was in receivership, agreed to fund it.[3]

At the next general meeting, in March 1874, all attention was given to the problems of land acquisition. One landowner at Pontesbury who owned a plot at the intended junction point with the Minsterley line was particularly intractable. The company gave up all hope, and revised their plans so that the junction would be made closer to Minsterley, but with all the works within the legal Limits of Deviation. This business aside, the directors hoped that all the necessary land would be acquired within the next six months. Progress was reported on the Bishop's Castle Railway link scheme.[4] Longueville's investigation concluded in August when they told More that they could not recommend the company to extend the railway as he would wish.

At a meeting chaired by Lord Bath, T. H. Lovett told More personally that his company should first complete its own line instead of 'merging it in a through scheme'. There were a number of important issues to influence the decision of the SDRCo. It had yet to raise the full authorised capital to enable its line to reach the terminus near Tankerville mine (see below). More's scheme, providing a potential route south out of the ore field, might have reduced their chances of raising this capital from other mine companies. It also might have reduced the potential of the SDR to make a profit. Being able to curtail the more efficient working of Snailbeach mine, his scheme again might have served better the other mineral Lords (which included More –

Pontesbury sidings, probably in 1936, in the most prosperous period of the Stephens group ownership. Shropshire County Council now controls its own road stone quarries in order to reduce costs, and as a result SDR traffic is high. The many large hopper wagons on the west siding show that the company is now using these rather than old coal wagons for carrying the County Council's Callow Hill stone. Meanwhile the narrow gauge to standard gauge transfer pier is still busy with Gravel Trading Co. and Whitstone Company deliveries, and the standard gauge siding is choked with wagons awaiting shunting or collection. On the east siding are a Union Railway Carriage & Wagon Co. coal wagon, two other open wagons and the explosives van. Visible in the right foreground, steadily grassing over, is the disused narrow gauge coal siding. H. F. G. DALSTON

In the late 1930s, *No.4* departs Pontesbury with a train of empty wagons.
R. E. TUSTIN
COURTESY P. INGHAM

he was landlord of Roman Gravels mine) rather than the Marquis of Bath. His Lordship had earlier agreed with the nascent SDRCo to oppose any competition (see Chapter Four). More's persuasion of the Midland Railway Carriage & Wagon Co. to stump up the cash for yet another competing (yet supposedly co-operative) scheme must have seemed a great threat.

Rejection of his scheme aggravated More, and he used his influence to create trouble for the company. In September 1874 the directors reported satisfactory terms arranged for buying the land for Railway No.1, but this was not completely true.[5] The Hope Edwardes family, which owned considerable land in the Pontesbury area, was still causing problems in January 1875, egged on by More. Longueville reported '... *correspondence with Mr. Jasper More who was apparently placing obstacles in the way of settlement.*' More eventually desisted after Hope Edwardes' solicitors were reminded of the company's powers under the Act. Despite all this, the directors had by September 1874 set Dennis to work staking out the line and preparing construction specifications for contractors. They were also due to begin the land acquisition for Railway No.2 and the sale of further shares to fund it.

Progress had been slow: the half-year's capital expenditure to the end of December 1874 had been a mere £59, whereas the total spent by the company to the end of June 1874 was £950. By 1875, the influential solicitor Theodore Martin, who lived at Bryntysilio, Llangollen, had joined the board. Specialising as a Parliamentary Agent, he made a good living from piloting private Bills for railways and industrial developments through Parliament. Martin was the Parliamentary Agent for the Glyn Valley Tramway, and owned quarry land in the Ceiriog valley which that line served. He was to be awarded the KCB in 1880 and the KCVO in 1896 by Queen Victoria for writing a five volume biography of Prince Albert. He became a personal friend of the Queen.[6] In February 1875 problems with land were continuing but the line of Railway No.1 had been marked out. In April the way looked clear for construction to begin. Advertisements were placed in the *Wrexham Advertiser and Guardian*, the *Oswestry Advertiser*, the *Manchester Examiner* and the *Shrewsbury Chronicle*. Several replies were received, but after some consideration the board accepted Elias Griffith's tender on 17 May. Elias Griffith was from Chirk, where he was a person of some note, being the proprietor of the Hand Hotel as well as a contractor, and was also Agent for the Chirk Castle Estate.[7] In the half-year to 30 June 1875, a further £2,000 of capital had been spent. Meanwhile Dennis took a fresh look at the line of the railway in the interest of efficiency. He found that by using curves of 2½ or three chains radius, the gradients could be eased to an average of 1 in 45. Many of the curves on the authorised plan were already as small as five chains radius. The Board of Trade,

Two views of *No.3* at Pontesbury in the late 1930s. COURTESY P. INGHAM

In about 1940, *No.2* with new smoke box stands at Snailbeach wharf. Clarence Rowson's old store now has a complete roof and is being used as a garage by the District Nurse.
H. F. G. DALSTON

which was then the regulatory body for railways, refused to allow such alterations, because it would have contravened the SDR Act (specifically Clause 14 of the Railway Clauses Consolidation Act 1845, which had been incorporated, and which prohibited such alterations after passing of the Act).

Henry Dennis wasted no time in getting Griffith on board so that work could start. It had been the best part of two years since the Act was passed, and the powers were limited to five years. Griffith and his two partners, Thomas Nuttall and John Davies Jones, were already working before their contract was signed on 1 August 1875. The first certificate for payment was made on 10 August. The construction contract, for the building of Railway No.1, consisted of thirteen numbered work specifications, each of which was expected to take a month. When satisfied, Dennis as engineer could certify each work package for payment on the last day of each calendar month. Thorough preparation allowed them to push through the first nine work packages in nine consecutive months, through to March 1876.

Elias Griffith and Henry Dennis already knew each other well. In May 1872, the same month that Dennis was appointed its engineer, Griffith was awarded the contract for building the Glyn Valley Tramway (GVT – see Chapter Three). Later he was a GVT director. Griffith was still owed a great deal of money by the GVTCo when he took the Snailbeach contract. Years after completion of the GVT, Griffith was still owed £740 out of a contract price of £9,552. At

No.3 photographed at Callow Hill on 19 December 1941. The shadows make clear how much the driving wheel tyres have been moved outwards on the wheel centres to re-gauge the locomotive.
J. G. VINCENT

Snailbeach, Dennis made sure that Griffith got his money promptly, and as can be seen from the foregoing, he had Griffith's trust.

As the construction commenced, so did the preparation for building Railway No.2. Although naturally to provide a greater source of revenue for the company, the line was expected to benefit Tankerville mine in the same way as for Snailbeach, by reducing costs. The Tankerville Mining Co. Ltd currently operated it.[8] Negotiations began concerning the partial funding by them of Railway No.2 by a share subscription. Henry Dennis presented the case for the railway side, in September 1875 attending a Tankerville Mining Co. board meeting to argue for a £3,500 subscription. The board was interested, but sought Counsel's opinion to see if the Articles of Association gave it the power to subscribe to a railway. Enthusiastic or not, the support was not forthcoming, and in March 1876 the railway company directors reported:

> ...The construction of Line No. 2 has been advisedly delayed, your Directors having failed hitherto in obtaining from the Mining and other interests which, locally, would be most accommodated by it, more than their good wishes, but without the subscriptions in substantial cash which they were, originally, led to expect. They have, therefore, recognised as their primary duty to the Proprietors the completion of Line No.1.

At the same time they had not abandoned the idea that Railway No.2 was worth building if the capital could be raised. The October 1875 report made no mention of Line No.2, but concentrated on the overall progress and the building of Railway No.1. At the end of June, a little over £3,000 had been spent in total, and by the end of the year this had more than doubled. More than £3,700 was spent on buying land, the construction work, engineering and surveying work and legal costs among other things. Elias Griffith's good work, as mentioned earlier, was to continue apace into 1876, and as the New Year advanced, the company was to have its answer from the Tankerville Mining Co. board: it would need to find its own capital for Railway No.2.

With the exception of the road underbridge at Pontesbury, all public highways crossed by the railway were maintained by the Condover District Highways Board. According to the Deposited Plans the crossings were to be on the level. The Railways Clauses Act of 1863 ensured that Dennis had planned for a crossing-keeper's hut at each of them. The same Act allowed the Board of Trade to order at any time that a bridge be built instead of a level crossing to ensure public safety. But provided Dennis followed the Act of Parliament for the line and the Railways Clauses Consolidation Act of 1845, he could build any unplanned bridges (and other types of construction) that he saw fit. Only bridges and tunnels appearing on the Deposited Plans needed to be built as planned. It was probably to avoid employing crossing-keepers and potential safety problems that Dennis built three bridges instead of level crossings. Late in 1875 the clerk to the Condover District Highways Board was told of the '... *interference with the Roads ... in constructing the Works ...*' of the SDR. In March 1876 the board surveyor met Elias Griffith, and reported back that the three bridges were unsatisfactory. Bridge and approach gradients were criticised. Only in October 1876 was the board satisfied with the various improvements promised by Dennis.[9]

By the spring of 1876 the decision had been made: Railway No.2 would not be built because of the difficulty in selling sufficient shares. Its estimated cost was about half the authorised capital of the company, but for a shorter line.[10] Only £13,900 of the total capital of £20,000 had been allotted, and this left the company with a problem: the borrowing powers given by the Act could not be used unless all the remaining shares had been taken up with the £2 deposit paid. At the May 1876 board meeting the position was explained, and it was resolved that the remaining shares be taken up by existing shareholders if agreeable, on the understanding that no calls would be made on these shares after the £2 deposits were paid. Following this, the Lovetts, John Jones Snr and Jnr and Henry Dennis took up an extra £2,000-worth of shares between them. At the mine company board meeting the following October, the railway company's position was noted, and further support given by nominating three Trustees to buy the remaining shares (£4,100) on behalf of the mine company using its own funds. In this way, all shares were allotted, but with only an extra £1,220 of capital being raised on a face value of £6,100.[11] The share issue now being complete, the railway company could take mortgages for up to £6,600. Shortly afterwards £4,000 was borrowed to complete the line.[12]

Looking at the amount of capital already spent, both Snailbeach boards made wise decisions. To the end of June 1876, a total of about £10,500 had been used, and only nine out of thirteen work packages for the completion of Railway No.1 had been paid for. The estimate for Railway No.1 was £9,534. Elias Griffith was working on package No.10 in June for prompt payment, and Pontesbury junction was not even past the planning stage. Railway No.1 was costing far above Dennis's budget, and looking at the rate of spending, it was arguable that all of the available capital would be needed for this line alone.

The winter of 1876 to 1877 was bad, and this combined with negotiations over the junction made progress difficult for Griffith. But by February 1877 the formation was mostly complete with more than half the track laid. Rails for most of the rest were ready for laying. By the end of April, track laying on the mine and smelt works branches was mostly complete.[13] The first locomotive *Belmont* had been bought, and negotiations were under way for the rolling stock. Two fitters from the Cambrian Railways delivered *Belmont* to Pontesbury on 7 June. Griffith moved *Belmont* to the junction by cart, and the fitters stayed for a few days to train the SDR's locomotive crew.[14] Griffith's certificate for package No.13 was signed on 23 June 1877, but the work was by no means finished. The rolling stock was delivered shortly after.

THE JUNCTION

Plans for Pontesbury junction were in circulation between Dennis, Jones and officials of the Great Western Railway (GWR) in October 1876. R. W. E. Johnston of the Great Western & London & North Western Joint Railway Engineer's Office in Birkenhead returned Dennis's plan with final corrections, telling him that the Joint Line's price for constructing their siding and retaining walls would be £500 excluding signals. Gong communication was needed in both directions between the Junction and Pontesbury Station. Signals were to be bought by the SDRCo from McKenzie Clunes & Holland. In early November

No.2 departs Pontesbury with a train of empties, probably June 1943.
L. W. PERKINS

Dickie Preece obligingly hauls each Baldwin locomotive in turn out of the shed for a photo parade on 11 June 1943.
The whistle of *No.2* is no longer the original one, now being mounted inside the cab. In the left background are later dressing floor sheds, while to the right is the crusher house.
L. W. PERKINS

Dennis instructed Johnston to proceed, and announced that the plans were agreed.

Meanwhile between October and December, John Jones was embroiled in a dispute with the Joint Railways Secretary and their Accountant's office. This concerned the junction maintenance and working agreement. Besides the construction cost, the SDRCo was to '... *sign an agreement undertaking to pay the wages of the signalmen; to defray the cost of removal, or the outlay of any additional works, which may be necessary should the Companies require to double the line, and to provide from time to time sufficient accommodation for working the traffic.*' Jones did not accept the idea that the company needing the junction was to pay for all the facilities, yet the Act of Parliament said so, but not in so many words.

Johnston's men set to work and finished their part of the junction by March 1877, and the GWR immediately applied for inspection.[15] This was well before the track laying was done on the narrow gauge side. Griffith was busy elsewhere on the line. In July the junction agreement was still not settled. The SDRCo still disagreed on paying for additional maintenance to the Joint Railway, but their secretary was adamant. '*It is only as regards the working of the Traffic that there can be any question as to the mode of dealing with the expense and we consider that our draft properly provides for this.*'

Dennis finally triumphed in 1878, when after a further volley of letters and some investigation, the Joint Railway made a concession. There would be no reduction in permanent way and signalling maintenance charges, but '... *as Mr. Patchett* [the branch line station master] *finds on further enquiry that the Porter at Pontesbury is not employed quite so long as at first appeared* ...' the annual expenses payable were reduced to £35.

With the agreement for working the junction under negotiation, at least some of the narrow gauge sidings and equipment laid at Pontesbury, and with a locomotive and rolling stock now on the line, it would at last be possible to accept traffic and bring in some revenue. This finally happened in July 1877, and a major part of Dennis's improvement plan for Snailbeach mine was essentially complete.

CHAPTER SIX

Prosperity and decline 1877–1900

The line opened for traffic in July 1877.[1] There was no inspection by the Board of Trade, as this was for lines intending to carry passengers.[2] It had already been decided that the line was to be initially for minerals and goods only (see Chapter Four). As a consequence of the railway opening, the Snailbeach Mine Co. and SDRCo director Major J. H. Lovett auctioned ten of his Snailbeach mine cart horses at the Smithfield market, Shrewsbury in early September.[3] What was a good thing for the mine company was a loss of private business for him. Although work continued that month at Pontesbury sidings, the directors reported '*... the Line so nearly completed as to enable the traffic to Snailbeach Mine to be taken over it ...*'[4] In October, Elias Griffith was relieved of his obligation to maintain the line for a year under guarantee, and of the building of two cabins, which were deleted from his work measurement No.13. He had lost money through the delay in raising funds.[5]

Dennis finally certified Railway No.1 complete in December.[6] The cost of construction (without Railway No.2) had already exceeded the estimate for the entire plan by over £500, and some payments were still due. In February 1878 profits were reported.[7] Net receipts in 1878 were £786, and £703 in 1879, resulting in dividends of three percent per year. Mineral tonnage was over 12,000 tons per year.[8]

In February 1880 the directors recommended that a second locomotive be bought, to be funded partly out of profits, so no dividend was paid that year. This was a measure of the seriousness of the motive power problem (see Chapter Eleven). In August 1881, the directors reported much improvement in revenue and that the new locomotive *Fernhill* was working.[9] In the following year a £500 debenture was issued, to pay the balance due on *Fernhill* and for repairs to *Belmont*. Mineral traffic increased by about 1,000 tons per year between 1881 and 1883 to over 16,000 tons, again allowing three percent dividends for three years running. These would be the last that the company paid. Cheap imports depressed the price of English lead to its lowest yet known, and the Snailbeach Mine Co. was forced to negotiate a royalty reduction from 1/10th to 1/16th in 1883.[10] The downturn in the mine's fortunes reflected on the railway. In 1883, railway operating costs were cut by dismissing the traffic manager and reducing wages, including putting the locomotive crew on short time.[11] Day-to-day running of the railway became another job for the mine managers.[12] Then in July 1884 the mine company decided to liquidate. Minerals carried halved to 8,433 tons in 1884 then slid to a low of 1,859 tons in 1886.

From the start, the SDRCo had relied on raising mortgages simply to finish construction. In 1884, despite an earlier agreement, the remaining share capital was subscribed, increasing it to the maximum allowance of £20,000. This allowed the mortgages previously totalling £5,500 to be reduced to £2,220.

From 1885, the mine was worked by the new Snailbeach Lead Mining Co. Ltd (SLMCo), which among other assets bought the 610 shares of the SDRCo held by the Snailbeach Mine Co. Ltd for £177 10s. The first directors of the SLMCo were Lt Col T. H. Lovett, Sir Theodore Martin KCB, Henry Dennis, C. W. Owen and C. W. Roberts, with another principal shareholder being John Jones of Mossfields.[13] Following the slump in traffic, Lord Bath agreed to halve the rent on the railway land to £11 per annum. Costs were minimised – for example in July 1888 almost all staff including the locomotive crew were working only a three day week. After August 1889, the SDRCo board stopped meeting, letting Dennis and the mine company carry the business on. Mineral traffic had recovered slowly after the slump, growing to 6,500 tons in 1890 while market prices improved, but traffic fell again. In 1891 a mortgage of £1,000 was recalled, and the mine company agreed to take on the debt. The mortgages of £2,220 at 4½ percent interest remained until 1899.

In 1888 Parliament approved the Railway and Canal Traffic Act, which was intended for 'the better regulation of Railway and Canal Traffic.'[14] In fact it eventually proved bothersome to the SDRCo. In accordance with this Act and with other public railway companies, the

This photograph of c.1885 by J. Thomas is one of the best photographs showing the central surface works of Snailbeach mine, and captures the scene during a slump in ore prices at the end of its most prosperous period, with all major developments in place. The various buildings and the railway sidings can be identified by referring to the plans and maps. A short length of the SDR main line, between its fences, can be seen to the left approaching what is now known as the Halvans bridge. It re-emerges from behind the white hillocks about half way across the picture. A little further distant is the smelt works with the furnace hall roof prominent, and to its right the branch line track bed climbing to the SDR line. To the immediate right of the white hillocks can be seen the old track bed of Eddy's smelt works tramway, leading from the ore house.

On the mine, the locomotive shed, main siding, carpenter's yard siding and various tramways are clearly shown with various tram tubs. In the foreground, the left hand tramway branch from the Landing Level tips onto the spoil tip tramway, which passes behind the head gear and the smiths' shops, while the right hand

branch tips onto the dressing floor. Following from left to right from behind the locomotive shed, the main railway siding can be seen passing behind a pair of disused waterwheel-driven buddles to a partly hidden slip road and 90 degree crossing, which join the siding with a dressing floor tramway, as described in the text. It then passes behind a revetted hillock holding a cistern on top, which can be seen to feed (or have fed) the waterwheel and dressing machinery using launders and pipes. Behind the revetment, the carpenter's yard siding diverges, and on it stands perhaps the line's entire stock of hopper wagons. At the end of the train is thought to be the miners' car, while a flat wagon stands on the main siding near the ore house.
The tramway which crosses the main siding divides on the spoil heap, a left hand branch proceeding onto a viaduct for tipping down a long chute into a crushing engine house. This later became known as the 'Halvans engine house', but it can be seen here long before the Halvans Co. Ltd began operating in 1900, and was mapped on the 1881 Ordnance Survey map. This journey was likely to be the next stage for dressing of an ore separated on the main dressing floor. NATIONAL LIBRARY OF WALES

company submitted to the Board of Trade in 1889 a revised classification of goods traffic and a revised schedule of maximum rates and charges. It did not have its proposal approved, and was finally sent a general schedule in 1891 to which it objected. The company's auditor, G. M. Jenkins, wrote the following to company secretary Edward Williams on 11 August 1891:

> I saw Mr Dennis on the matter this morning & pointed out to him that with one exception the proposed alterations in the Rates will not affect the S.D.R. to any material extent.
>
> The alteration in the Rates to which exception is taken is the one referring to the schedule under the different classes whereby the present rate is considerably reduced.
>
> Mr Dennis wishes the Board of Trade notified that this contemplated alteration is objected to for the following reasons.
>
> 1. Having regard to the fact that the Railway is only 3¼ miles long & the benefit bestowed on the local inhabitants by the service afforded it is unfair to the company to impose such a small charge per mile, under the different classes referred to & that the company should be protected by a minimum charge as to distance. For instance in the case of large companies who can take advantage of the distance of 20 miles the reduction would not be so noticeable as in the case of the S.D.R. whose length of line is only 3¼ miles, & this should be taken into consideration.
> 2. That the present rates give satisfaction to the public & are in no way objected to.
> 3. That the line at the present time is worked at a loss.
> 4. That the proposed reduction would mean a still further loss to the Company which unfortunately would inevitably lead to the line being closed.
>
> Under these circumstances it is therefore asked that no alterations be made in the present rates per mile which are not in any way excessive considering the length of the line & the object it serves.[15]

The Board of Trade's subsequent Confirmation Order, made legal by the *Railway Rates and Charges No.14* Order Confirmation Act 1892, grouped the SDR with the Midland and South Western Junction Railway Co. and eight others. The SDRCo did not get what it wanted, and by some glaring omission failed to apply the new prices for another seven years. This had financial repercussions, as will be described later.

Coincidentally with rising prices and improvement in demand for barytes, a small barytes milling business was begun at Snailbeach in 1889, although it had little or no impact on the fortunes of the railway. Alfred Haywood of Bayswater, London, built a corrugated iron mill building on a waste piece of ground on the east side of the railway, opposite the White Hillocks. Here he installed a steam-driven grinding mill and traded in ground barytes. A lease for the mill was completed with the SLMCo in December 1890. The business was quickly bought by the South Shropshire Barytes & Lead Co. Ltd, which employed Haywood as a manager. Haywood and the successor company used the railway for bringing in stores and despatching ground barytes. A further company called the Barytes & Lead Co. Ltd bought the business, but was in liquidation by 1900. It is not known if it used the railway.[16]

In 1892 the price of lead was dropping fast, affecting sales and railway traffic, while the price of coal was rising. The SLMCo mortgaged its shares in the railway to the bankers Croxon & Co. to raise funds, and began negotiations with Lord Bath to find a way through this trouble. It was made clear to Lord Bath that without concessions, there would be no alternative other than closing the mine, and the railway with it. His Lordship agreed a three year suspension of royalties; a temporary suspension of smelting until the price of coal fell enough to allow a profit to be made; and allowed the sale of ore. Earl Tankerville agreed similar conditions for the Vessons mine, which was in effect an extension of the Snailbeach mine workings eastwards under his land. In fact the price of coal, like the costs of industrial production generally, had begun to inflate, and regular smelting was never resumed. The mine company was fighting to maintain sales in lead and zinc ores, and was increasingly relying on barytes for which the carriage rate per ton was lower.[17] For the rest of the decade, traffic on the railway settled at around 5,000 tons a year. There were a few loss making years, but a small annual surplus was usually made.

With the renewed problems of 1892 came the wind of change. In that year, T. H. Lovett, J. H. Lovett and John Jones sold their shares in the SLMCo to Henry Dennis, his son Henry Dyke Dennis and F. B. Owen. This ended over a century of Lovett family involvement in Snailbeach mine. The position had become so bad that they saw no future in it. The old order continued with the SDRCo, but it was not to last much longer. The railway was utterly at the mercy of the mine company – so much so that between August 1889 and May 1893, no separate board or shareholder meetings were held by the railway company. Being left in charge, the mine company was keen to cut costs as well as borrowing to get by. It had withheld payments to the railway company on the basis of being owed interest on the SDR's mortgage. The situation came to a head in early 1893, and in May the two boards of directors agreed formally that the line would continue to work as it had done for some time, with the mine company servicing the railway's debts and balancing the books. As well as having mortgages, the line traded at a loss in 1891 and 1893. The mine company told Lord Bath that the railway company was unable to continue operating without its help. The only way it might have paid its debts would have been through liquidation, but this would have left the mine without a transport system. This agreement to continue working the railway while it remained in debt to the SLMCo marks the beginning of a gradual move to total control by the mine company. In July 1893, with the knowledge of the railway company board, the May agreement was ratified when the mine company resolved that '... *The Mine will continue paying the cost of Working the Railway and the Interest on their Debenture Stock, taking the receipts and keeping a strict account of the Receipts and Payments* ...' In September, the mine company asked Lord Bath to forego the rent on the railway land for the time that it kept the railway open. Lord Bath agreed, provided the rents on the mining land and cottages were paid regularly.

The SLMCo Board, which was dominated by the Dennises, though they held little more than 40 percent of the shares, inched closer to total control of the railway. Trade remained poor and temporary closures were made.[18] In December 1894, an offer of five shillings per £10 share was made to all SDR shareholders, effectively valuing the company at £500. There was too little response and no shares were bought. From 1894 to 1898, there were modest traffic revenues, but expenses were mounting, interest on the debentures stayed in arrears and the railway company was in continuous debt to the mine company. In May 1898, Dennis warned the SLMCo board that repairs to the locomotive *Belmont* costing up to £200 were needed, since they would be paying for it. The question of take-over became hot. The mine company prepared its accounts in evidence, and in August 1898, Edward Williams Vaughan, the railway company secretary, agreed to '... *recommend the shareholders of the Rys Co to transfer the Ry to the Mining Co on the condition that the latter Co. guarantees the Debentures of the Rys Co* ...' In the autumn the matter was deferred again, and the mortgage interest up to 1897 was ordered to be paid. Then in January 1899, Lord Bath found that the mine company had broken the rental waiver agreement, and asked the company to pay 7½ years' worth of unpaid rent on the railway at £22 per year. This demand led to decisive action. A tender was sought for cart transport between Minsterley and the mines, and in February an offer was received from Snailbeach haulier Percy Rowson for 1s 4d per ton down to Minsterley and 2s 4d per ton up, which seemed less than railway costs. The railway company was then told that if Lord Bath would forego the rent, the mine company would be willing to take over the railway entirely, including its liabilities. Failing such an agreement, a carting contract would be made, which would effectively seal the fate of the railway.

The take-over was under way. In June 1899 Lord Bath agreed to sell his railway shares to the mine company at 5s each. The SLMCo was planning to spend £1,500 on refurbishing the line, and in consideration of this, his Lordship also offered to waive the rent arrears and reduce the railway rent to £5 per year. With this news, a decision on a switch to carting was postponed. All other shareholders were approached, and by September the mine company owned more than three-quarters of the SDRCo shares.

By December 1899 the new SDRCo board of

A postcard view of Snailbeach before the Great War. The photographer is close to the Halvans bridge. In the foreground is the barytes mill built by Alfred Haywood around 1890. He installed a five feet diameter globe mill driven by a semi-portable steam engine to grind his product. Haywood used the railway but there was no siding, so stores and barytes were unloaded and loaded with the wagons left on the line, which is just out of picture to the right. With the locomotive depot being above the mill, it is likely that Haywood's wagons were run down to Pontesbury by gravity.
COURTESY E. ROWSON

Henry Dennis, H. Dyke Dennis, F. B. Owen and William Edwards was in place, with Henry Dennis elected chairman. From now on, both the railway and mine company boards would be the same. The line made an operating loss of £110 for the year. The mine company's Articles of Association were amended to allow the directors to invest profits in SDR shares and debentures, and it continued to pay the railway wages. A balance sheet showing the condition of the accounts between the companies from 1892 to 1899 was then ordered. The board was also reminded that *Belmont*, having been under repair, was held at Cudworth & Johnson's works with the bill unpaid. Auditor G. M. Jenkins reported that because the company had failed to implement revised rates and classification for carriage of goods traffic following the Railway and Canal Traffic Act 1888 and the Board of Trade's subsequent Order, there was a balance in favour of the mine company because of under and overcharging. In total, the mine company was owed £625 81s 4d and Cudworth & Johnson £340 1s 2d – the latter being rather more than Dennis's earlier estimate. In April 1900, at an extraordinary general meeting called to discuss the position, the railway company shareholders were told that the Statute of Limitations applied to the mine company debt. Only six years' worth of charges prior to the company admitting liability were payable. This debt was ordered to be paid with interest, totalling £640 10s, and liability admitted for Cudworth & Johnson's bill. At the request of this meeting, the SLMCo agreed to continue funding the excess of expenses over traffic revenue. Another loan was made to cover the SDRCo's debt including the repairs to *Belmont*. Total expense on locomotive power of £529 in 1900 was a prime cause of the year's net operating loss of £464. The company now had mortgages of £3,201.

By the turn of the twentieth century, the SDR had declined from being a profitable entity to one which depended entirely on subsidy for survival. The mine company could have switched to carting easily – traffic was low enough that contractors could have coped. Instead the Board decided to service the railway's losses out of the mine's profits and keep the line open. The Marquis of Bath helped by offering concessions in cash to both companies from his expected income. There are three possible reasons for this happening. The first (although nowhere is it stated directly) is that the railway was viewed as essential for effective operation of the mine and transport of the produce. Around this time the mine company was still exploring for new mineral reserves at Snailbeach, and also making judicious investments in machinery to improve production economy.[19] Closing the railway would not have made sense. Secondly the line provided employment to Bath's tenants, many of whom were also mine company employees – casual as well as full time. Thirdly it was a valuable public service, as was stated by G. M. Jenkins. In 1891, for example, the Carriage Accounts book names many members of the public who had goods carried for a charge of a shilling or two, or even pennies. The mine company would not have been alone in needing to switch to carts. The public already had this option, but there were advantages in using the railway, or they would have switched already. A cynical view might be that his Lordship was making only small concessions, with the simple aim of maintaining his royalties. Yet closure of the railway would have resulted in a carting contract, with no real effect on his royalties at all, other than the railway rent.

The SLMCo persuaded the Marquis of Bath to waive the arrears of rent (which included unpaid cottage rents), to further reduce the railway rent and to give up his shareholder influence, when he had least to lose by allowing the railway to close. Since the boom years, jobs in the south west Shropshire mines had dwindled and the pay decreased. People were drifting away to the cities and towns for more secure work as the district became poorer. In this atmosphere, the mine company directors convinced his Lordship to share some of the loss with them. This was to help keep the community alive, as well as to maintain a more effective transport system than carters could provide.

EXTENSION AND TAKEOVER PLANS AND OTHER SCHEMES

As mentioned earlier in this Chapter, the price of lead plummeted in the early 1880s, but in1884 to 1885 there was an upturn, followed by a modest general improvement in price until 1890. Demand for barytes was also increasing. This led to some new investment and output in south west Shropshire mines, and a new small crop of speculative mineral railway schemes. But prices fell again to an all-time low in 1894, killing off any such big ideas which still had breath. Only the Light Railways Act of 1896 was to hold any hope of new railway development in the district.

Hulton Harrop's extension scheme

In 1885, Mr W. Hulton Harrop of Lythwood Hall, Shrewsbury made enquiries regarding an extension to the Snailbeach railway. Hulton Harrop owned a large estate on the east side of the Stiperstones, where the increasingly popular mineral barytes was found in valuable pockets. The area was and is isolated, being served only by narrow lanes which were far from suitable for mineral traffic. Henry Dennis prepared an estimate for the line, which was to include a 200 yard tunnel through the Stiperstones ridge and take a route through the Gatten onto a '... *low summit above Bog Mine ...*' The length of the extension was to be a little short of five miles. The cost of construction was estimated at £13,451 15s 5d, calculated at a rate of £1,847 6s 0d per mile and the tunnel being excavated for £3,400.

Although this scheme came to nought, Hulton Harrop opened a barytes mine on his estate at Gatten a little later. In 1887 to 1890, a recorded output of only 1,713 tons of barytes worth about £1,132 was raised there. The Bog mine had stopped work in 1884 along with Tankerville, when its parent company the Tankerville Great Consols Co. Ltd liquidated. Although no proof has been found of the idea, it is possible that Hulton Harrop was considering re-opening the Bog mine as well as starting his own mine, resulting in the railway extension scheme described. Considering the estimated cost when compared to the output of Gatten mine, it is hardly a surprise that it was never built.

The Shropshire Mineral (Light) Railway

In 1890, a Bill was being prepared for Parliament to authorise a 2ft 4in gauge railway making a junction with the SDR at Snailbeach. The named company was George H. Andrewes, William Jallings, Frederick Eddles and Thomas J. Atkinson, and the engineer was George W. Usill. The route was long and difficult for both construction and operation, heading roughly south via Tankerville and Pennerley to the Bog and Rock mines, there to turn northward onto the east flank of the Stiperstones to Gatten. A branch line would serve mines in the Hope valley including Roman Gravels.[20]

On 18 December 1890, W. Schofield, a railway contractor of Bucklersbury, London, made a written offer to Longueville of £14,000 for purchase of the SDR, subject to the Shropshire Mineral (Light) Railway Bill

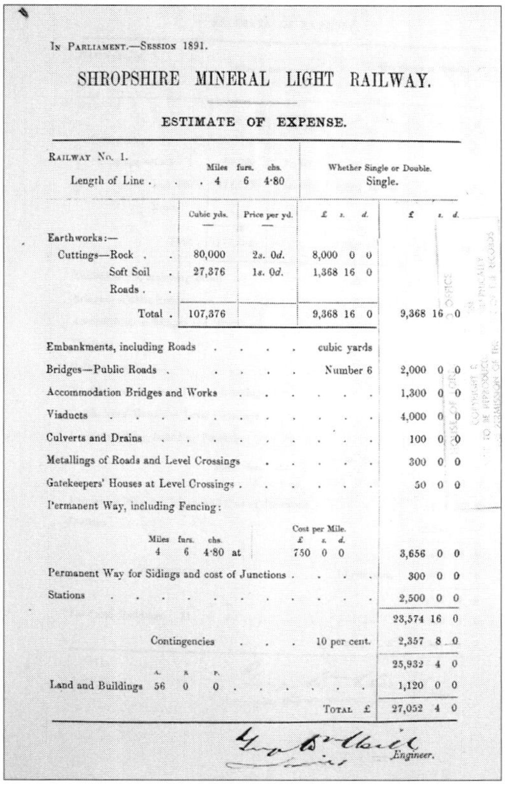
IN PARLIAMENT.—SESSION 1891.

SHROPSHIRE MINERAL LIGHT RAILWAY.

ESTIMATE OF EXPENSE.

RAILWAY No. 1.

Length of Line . . . 4 Miles 6 furs. 4·80 chs. — Whether Single or Double. Single.

	Cubic yds.	Price per yd.	£ s. d.	£ s. d.
Earthworks:—				
Cuttings—Rock . .	80,000	2s. 0d.	8,000 0 0	
Soft Soil .	27,376	1s. 0d.	1,368 16 0	
Roads . .				
Total .	107,376		9,368 16 0	9,368 16 0
Embankments, including Roads			cubic yards	
Bridges—Public Roads			Number 6	2,000 0 0
Accommodation Bridges and Works				1,300 0 0
Viaducts				4,000 0 0
Culverts and Drains				100 0 0
Metallings of Roads and Level Crossings				300 0 0
Gatekeepers' Houses at Level Crossings				50 0 0
Permanent Way, including Fencing:				
Miles 4, furs. 6, chs. 4·80 at			Cost per Mile £750 0s. 0d.	3,656 0 0
Permanent Way for Sidings and cost of Junctions				300 0 0
Stations				2,500 0 0
				23,574 16 0
Contingencies . . .			10 per cent.	2,357 8 0
				25,932 4 0
Land and Buildings A. 56 R. 0 P. 0				1,120 0 0
			TOTAL £	27,052 4 0

Engineer.

Extract of the Shropshire Mineral (Light) Railway Act – Engineer's Estimate of Expense.
PARLIAMENTARY ARCHIVE

becoming law. A little later, a draft agreement for the sale and purchase of the SDR line, lands, rolling stock and plant by the Shropshire Mineral (Light) Railway Co. (SMLRCo) was made, between Schofield and Edward Williams as SDRCo trustee. The SMLRCo was incorporated by Act of Parliament in August 1891 to build the line and have running powers over the SDR. Share capital of £75,000 with mortgages up to £25,000 were authorised. It was to be a 'light railway' as defined in the Regulation of Railways Act 1868. In late 1891 Henry Dennis, Hulton

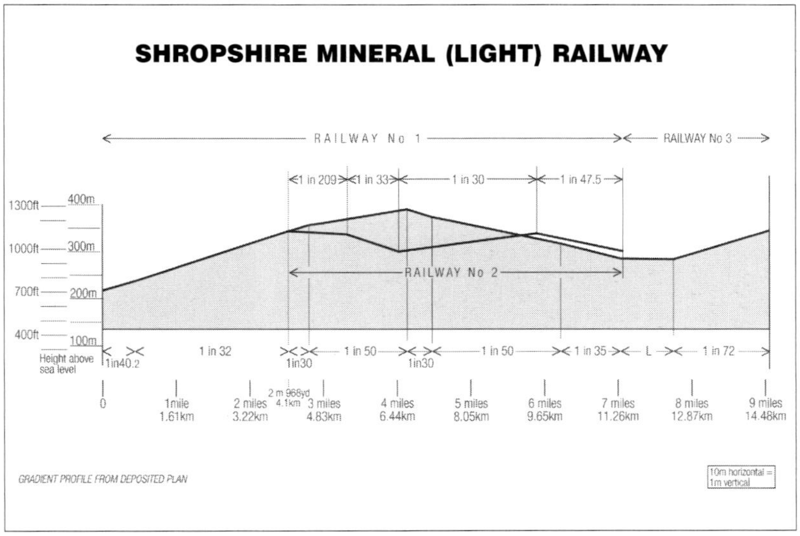

Gradient profiles of the Shropshire Mineral (Light) Railway.

Map of SDR and south Shropshire railway schemes 1890 - 1897. Hulton-Harrop's scheme of 1885 has not been indicated, as a useful route with only a 200 yard tunnel was impossible to guess at.

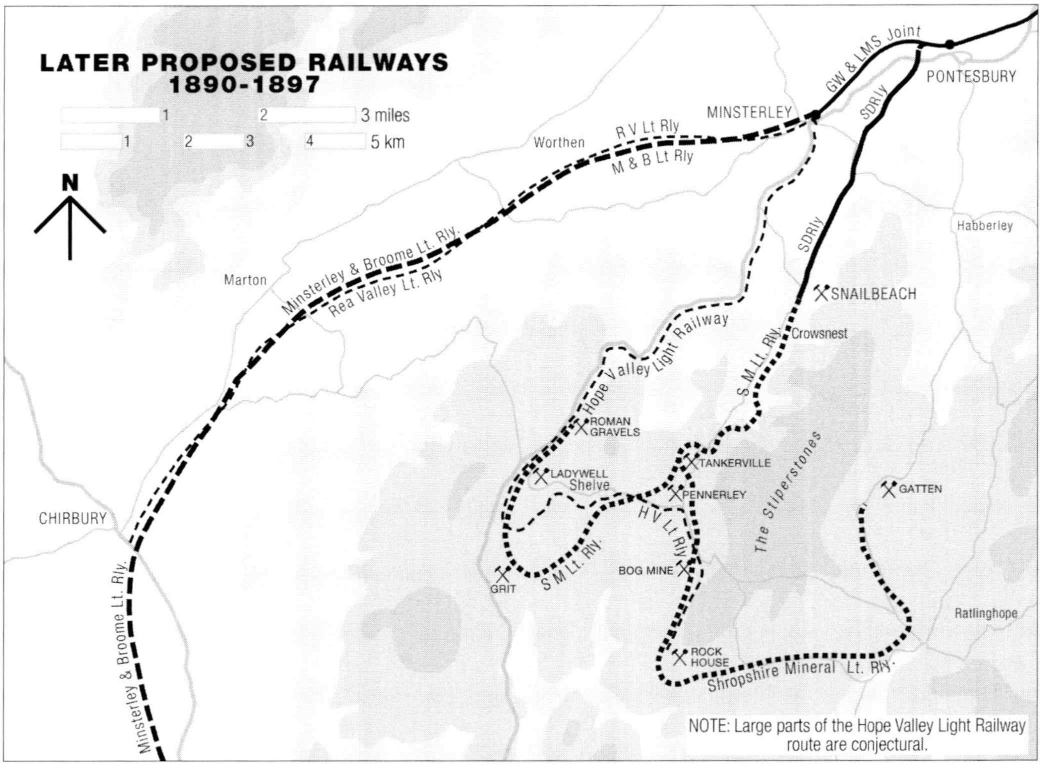

Harrop, R. Jasper More MP, Edward Williams and Viscount Weymouth were approved by Dennis to replace the first authorised directors, with the comment '*... but parties finding money should have a Director say in place of Mr More & that Lord Bath should be asked.*' Dennis also confirmed that the SDRCo would need an Act of Parliament to sell the railway outright. Hulton Harrop, More and Viscount Weymouth were major land and mineral lords (the latter being heir to the Marquis of Bath), whose land the combined lines would pass through.

Mineral prices slumped in 1892, but in 1894 the SDRCo made a second sale agreement, this time with the National Mercantile Investment Co. Ltd acting for the SMLRCo. This was the scheme's last gasp – no capital was raised, and the company's powers lapsed with nothing being built.[21] Of a similar character and cost to Kough's second South Shropshire Railways scheme, it too would have been a white elephant. When the SMLR plan was born, prices had risen and there was greater demand for minerals. But even the major mines in the field, all to be served by this line, were past their best. Roman Gravels, like Snailbeach, worked through the 1880s but was in decline. By 1889 it was raising less than 1,000 tons of lead ore with 150 to 200 tons of zinc ore per year. By 1895 it was not working. In response to the higher prices, the Tankerville, Pennerley and Bog mines were re-opened by their landlords in 1889, but all with poor results. The output in 1890 of Gatten, the newest enterprise to be served, was barytes worth only £574. The price of lead then fell again to under £10 a ton by 1894, making almost all the mines uneconomic.[22] No-one in their right mind would have lent or borrowed money to build this railway. The Shropshire Mineral (Light) Railway scheme was dead in the water.

John Job's Hope Valley light railway or tramway scheme

As early as 1894, the SLMCo's agent John Job had done some work on getting approval for a light railway route south from Minsterley via Plox Green and The Rock. Minutes of meetings of the Condover Highways Board and the Shropshire County Council (SCC) Roads and Bridges Committee mention this scheme.

At a meeting of the Condover Highways Board on 8 September, the board surveyor was instructed to see Mr Job to find out exactly what he wanted.[23] Then a letter dated 12 October 1894 from the secretary of the Stiperstones Quarries Co. to the SCC Roads and Bridges Committee outlined a proposal for the line. Details of this letter minuted at the Committee meeting of 13 October show that it was to have '*... rails of narrow guage* [sic], *similar to Town Street or Road tramways,* *on the side of and along the Main Road leading from Minsterley to Plox Green and other places ...'* A special committee was to be appointed to investigate the proposal and report to the SCC.[24]

At the Condover Highways Board meeting also held on 13 October 1894, which considered John Job's proposal, it was noted that if the SCC would agree to the light railway between Plox Green and Minsterley station, the Condover Highways Board would agree a continuation from Plox Green to The Rock. The minutes of the board meeting of 10 November 1894 note that '... *The clerk was instructed to write to Mr Job and inform him that the matter of his application to lay a light railway is remaining in abeyance until the County Council had come to a decision upon the application made to them ...'*

No other information has been found concerning this proposal. Such a scheme, in competition with the SDR, would not have been approved by the SLMCo Board. This became evident with further schemes John Job was involved with, and it must be presumed that he was pursuing this idea privately, in an attempt to get a railway which would serve various mines and quarries on the Stiperstones, in the Hope Valley and further south, now that it was obvious that the SMLR was not to be built. Once more, needless to say, nothing ever came to pass.

The Minsterley and Broome and Rea Valley Light Railway schemes

After Robert Jasper More's last chance for a mineral railway died in 1894, the Light Railways Act of 1896 provided him with fresh impetus for promoting railways. Approved schemes under the Act were expected to improve agriculture. Funds in the form of loans, and free grants in exceptional cases, were available from the public purse. To qualify for loans, the promoters first had to raise shares for half of the capital needed. Half of these shares were to be fully paid up and subscribed by bodies other than local authorities.[25] In 1895 the preceding Light Railways Bill was being prepared for Parliament, and in November the Parliamentary Committee staged a Light Railways Conference in London to promote it to interested parties. Attendees included the railway engineer and promoter Holman F. Stephens (see Chapter Seven). More was invited but did not attend.[26]

R. Jasper More was a chief proponent of the Minsterley and Broome Light Railway, as with earlier schemes. In correspondence with More in May 1896, Henry Dennis was agreeable that the Joint Railway companies (the Great Western and London & North Western Railway companies, who jointly owned the Shrewsbury & Welshpool lines – see Chapter Four) might take over and extend the SDR under the provisions of the Light Railways Act, then being debated in Parliament. The Act became law in August, and More wrote again about raising capital to '... *make a line from Minsterley to Lydham ...'* using another north-south route. A prospectus for the Minsterley and Broome Light Railway via Bishop's Castle and Clun was published by a steering committee including Snailbeach mine agent and accountant John Job. It is possible that John Job still advocated the Hope Valley route. Some members of the committee favoured 'the Main Road route', meaning via Longden and Pulverbatch to Bishop's Castle. (This was a turnpike road before the Shrewsbury, Minsterley and Bishop's Castle turnpike roads – now the A488 – were built. It was later a first class road and is now unclassified.) The SDRCo did not support this plan, but remained sympathetic to take-over and extension plans. The committee asked for land to be given free, and promoted shares at £1 each. Then a deputation led by the Mayor of Bishop's Castle met the Joint Railway companies to discuss how the latter might assist with the scheme, but nothing came of it and no plans were deposited. With no support the idea fizzled out like a damp squib, but years later More was still enthusiastic about extending the Bishop's Castle Railway. In a letter printed in the *Shrewsbury Chronicle* of 30 August 1901, he made a whole raft of suggestions about the viability of such a scheme and its potential for opening up the area.

Traction engine and trailer of Shropshire Lead Mines Ltd. The engines hauling up and down the lane between Plox Green and the Bog mine, to and from the branch line at Minsterley, were competitors of the SDR. They also played havoc with the unmetalled road. The ruts their wheels made were so big that it is said a person could lay down unseen in one. Such damage made the roads practically unusable for other hauliers as well as the public, and they needed extensive repairs at the expense of ratepayers, which included mine companies. This led to the building of the aerial ropeway by Shropshire Mines Ltd, and even less traffic for the SDR.
COURTESY P. FRANCIS

The more modest Rea Valley Light Railway plan was also afoot in the late 1890s, although it had nothing to offer the SDRCo. Promoted by the Chirbury Rural District Council, this scheme was another repeat of the Minsterley to Chirbury route first mooted in the 1850s (see Chapter Four). Plans were deposited in May 1897.[27] An application for a Light Railway Order was made to the Commissioners shortly afterwards, but again it came to nothing.[28] A final proposal to extend the Minsterley branch to the Bishop's Castle Railway was made in the 1920s.[29]

A thread running through these various schemes is the positive reaction of the SDRCo to take-over. The company was never likely to be able to pay its way out of debt (let alone pay dividends), and to receive an offer of £14,000 for the railway was a golden opportunity. But caution finally prevailed with the incoming investors and the SDRCo was left to struggle on, until its dependence on the SLMCo became absolute.

A vision of road locomotive haulage

In the midst of Jasper More's campaign for the Minsterley and Broome line, Samuel Harley Kough (see Chapter Four) wrote to the *Wellington Journal* on 5 September 1896, ostensibly on the subject of light railways. He had become a far wiser man in the space of twenty-five years. Instead of a new railway south of Snailbeach to Bishop's Castle, he proposed that a farm road from the Green, Perkin's Beach (now known as Stiperstones), be widened and metalled and a bridge built over the Hope brook. This would improve transport for several mines. He said that his proposal anticipated enforcement of the recent Locomotives on Highways Act. He added: '*... Be not beguiled by any chimera of extending the Snailbeach Narrow Gauge Railway to Lydham Heath ...*' Kough's road was not built, and later, steam road locomotives with trailers rumbled through Snailbeach and Stiperstones along the old lane.

Kough was wrong about the scope of the Act he mentioned. The Locomotives on Highways Act 1896 recognised the existence of 'light locomotives' which might carry people and use petroleum as fuel. They would be permitted to travel on public roads at speeds up to fourteen miles per hour. In short, the motor car was fast becoming a reality. And they needed to be taxed and regulated! In fact it was the Locomotives Act of 1898 which recognised the widespread and increasing use of road locomotives for haulage. County and Borough Councils were obliged to issue licences for them, and were empowered to create byelaws to regulate their use. The byelaws followed after the fact, and took much time and effort to implement. In the Edwardian era, road locomotives grew in popularity with mineral hauliers in Shropshire – including those in the Hope valley – but they made enemies for themselves (see Chapter Seven). The steam road locomotive with trailers was only suitable for short haul use, and was killed off within a quarter century by rapid advances in motor transport. Kough's vision of new roads for freight traffic would have to wait – for the long haul and for the age of the petrol and diesel lorry.

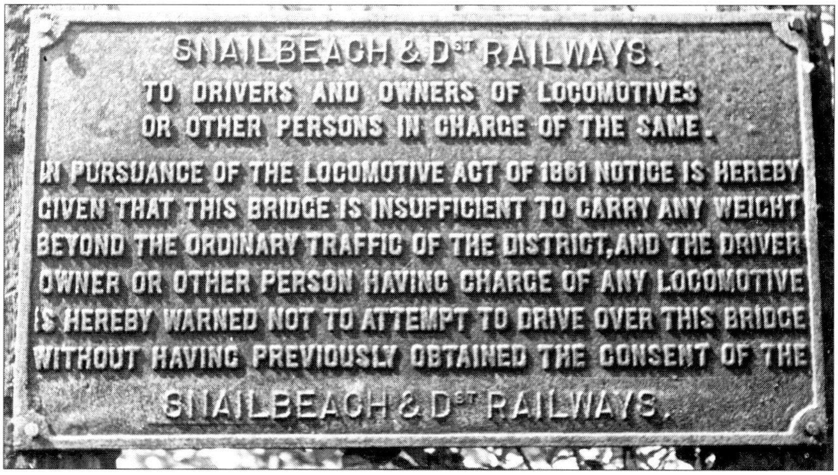

Bridge warning notice for users of road locomotives. The two bridges in Snailbeach were regularly used by locomotives and trailers going to and from Bog mine. This notice has survived, and at the time of writing is on display at the Winchcombe Railway Museum & Garden, operated by Gloucestershire County Council

CHAPTER SEVEN

*'For the benefit of the District'**
1901–1922

* Arthur H. Bardswell, 6 Oct 1916 (NLW Longueville letter books)

Under a common board of directors, the practice of the SLMCo granting new debentures to the SDRCo to cover the unpaid interest continued between 1901 and 1907. In 1908 the railway company exhausted its authorised borrowing limit of £6,300, while in most years the operation lost money. Despite a new company, the Halvans Co. Ltd (managing director William Kinross Arber) taking a lease to work the Snailbeach spoil heaps in 1900, there was little change in tonnages hauled. Only about 5,000 tons of minerals per year were carried until 1904.[1] The Halvans Co. was well managed, and having compared Glyn Valley Tramway carriage rates, immediately negotiated a rebate of threepence on every ton above the first 70 tons carried per calendar month. In 1902, it was granted a rate of one shilling per ton for spar carried subject to a minimum of 150 tons per month.[2] The Halvans Co. either carted the minerals and coal or used the railway depending on costs.[3] The spar it extracted from the spoil was principally used as building material.

Besides the new sub-lease to the Halvans Co., further sub-leases were made in the early 1900s of the Snailbeach barytes mill built by Alfred Haywood. As before, they created very little traffic for the railway. In 1901, Mr J. W. Walker of London EC leased the mill. Walker was a tenant until 1904, but it is not known if he used the railway. Then a Mr F. W. Read, trading as the Snailbeach Barytes Co., took a lease of the mill and worked there from 1906 until 1911. Read was a small user of the railway.[4]

In 1905, the railway company gained an important new customer, the Granham's Moor Quarry Co. Ltd (GMQCo), with offices at 5 The Square, Shrewsbury.[5] This operation began as the partnership William Toye & Co., consisting of Toye, an Inspector of Explosives, and Thomas G. Marsh. They took a lease of Nill's Hill quarry from John Humphreys in 1901, then in January 1904 leased Granham's Moor quarry in Eastridge Wood from George Montagu Earl Tankerville.[6] Silica stone from the latter quarry was used as road metal by the Shropshire County Council (SCC) as early as 1900.[7] Shipping of this stone via the railway began in 1905, but exactly how the line was reached then is unknown. Toye & Co. turned to the railway after protests about the use of steam road locomotives (traction engines) for hauling stone between the quarry and

A pre-Great War postcard of Snailbeach mine. On the left can just be seen a locomotive, thought to be either *Fernhill* or *Dennis*, hauling empty wagons out of the coal yard. To the left of the loco, other wagons wait at the bottom of the incline. A small area of the picture near to George's Shaft headgear is unfortunately spoilt, but including the view was essential because the train movement shown is unique in that era. COURTESY E. ROWSON

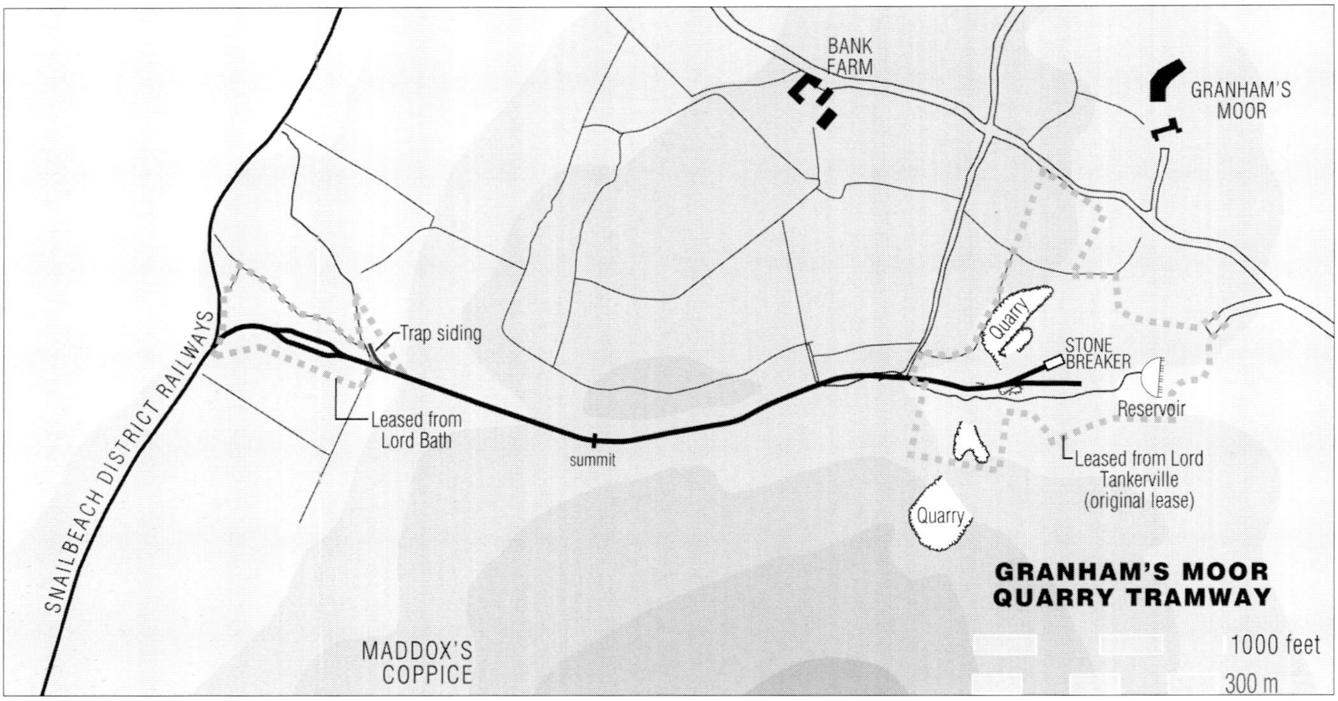

Map showing the extents of the Granham's Moor quarry tramway lease from the Marquess of Bath and the quarry lease from Earl Tankerville. The route of the Tramway was also leased from Lord Bath as far as Lord Tankerville's property.
SA 2079/40.14;
SA 4743/50;
SA 4011/26;
OS SJ30SE 1954 6in

Pontesbury. The Atcham Rural District Council and 46 ratepayers and users of this road asked the SCC to restrict locomotive traffic on it by their powers under the Locomotives Act 1898. But Toye had already informed the SCC Roads and Bridges Committee of 'certain arrangements' made to take the engines off the road, so no action was needed. Traction engines were an increasing nuisance generally. By 1906, some Councils in Shropshire were applying for byelaws and time restrictions to control them.[8] In this case, the operator was forced to look for alternative transport.

Toye & Co. not only found a possible solution to their transport predicament on approaching the SDRCo, but also a source of new capital. Henry Dennis, his sons Henry Dyke and Walter Pen, and William Edwards formed a new partnership with Toye and Marsh, which bought the quarry business and registered as the

Postcard showing the Granham's Moor Quarry Co. tramway. Photographed from a point above the wagon, the haulage rope and guide, telephone wires and a trap for runaway wagons can be seen. The bank of stone rubble and chips on which the trap siding was laid can still be found just inside the wood.
COURTESY I. J. BROWN

Showing un-ballasted track and a pristine cutting, this and the previous photograph must date from 1906 at the latest. The workmen are posing on a new-style hopper wagon with steel dumb buffers, held on the haulage rope. The cutting was in the field between the wood and the SDR junction. It has been completely removed, and today there is no visible sign of railway sidings. COURTESY I. J. BROWN

Granham's Moor Quarry Co. Ltd on 29 July 1905. Toye and Marsh retained a majority share holding. The new company's assets included '... *the Lease procured or to be procured from the Marquis of Bath for surface rights of way.'* This allowed the stone to be hauled over Lord Bath's land between the quarry and the SDR. On 25 August 1905, the Marquis of Bath granted a lease to the quarry company for a tramway through Maddox Coppice to the SDR.[9] The quarry company also took a new 21 year lease of Granham's Moor quarry dated 29 September 1905.[10] Equipment at the quarry included steam-driven crushing machinery, a tarring plant and internal quarry tramways, and a reservoir was built.[11] Coal for the machinery was brought in by the railway.

When the GMQCo was formed, the SDR locomotives were not in very reliable condition. Higher tonnages were anticipated with the new quarry traffic and reliability would be needed. Henry Dennis arranged that the Glyn Valley Tramway Co send its locomotive *Sir Theodore* for

The crushing, screening and tarring works in Granham's Moor quarry on a working day. Besides the smoke and steam which show this, there are three workmen posing in the shed doorway. But the family in the foreground is in best clothes. Note the side-tipping railway wagon to the right of the shed. The quarry buildings and surroundings bear no resemblance to the extensive concrete remains found in the quarry today. This view is thought to show the quarry as developed by Toye & Co. with the tramway by the Dennises, while today's remains are those of the GMQCo under the ownership of Gerald Briggs. COURTESY I. J. BROWN

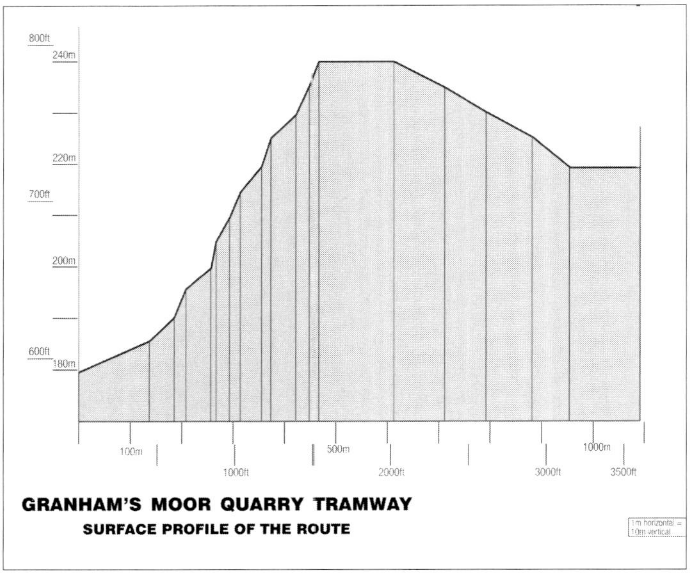

Surface gradient profile of the Granham's Moor quarry tramway route. The gradients show clearly why some means of haulage other than an adhesion railway was needed.

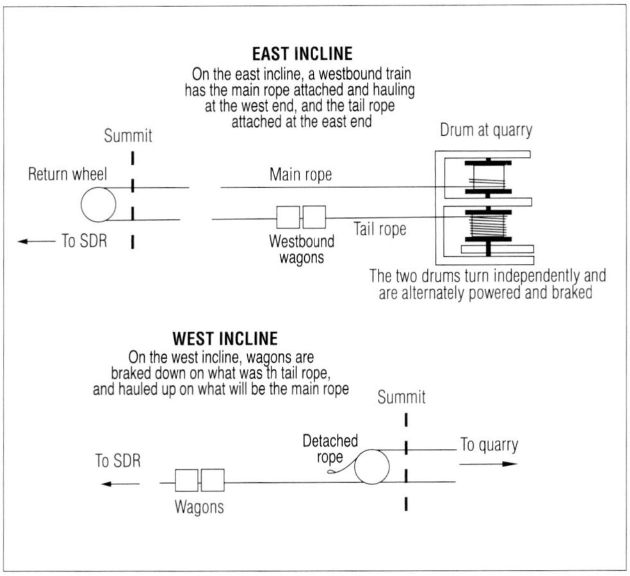

Schematic diagram to show probable means of main and tail haulage on the quarry tramway. Main and tail haulage driven from the quarry would account for the return wheel and obvious lack of an engine house at the summit. The winding engine at the quarry might have had an indicator, like those on mine winding engines, showing the position of the train (as with a cage or bucket in a mine shaft). Using this, the engineman could with practice stop the train correctly at the western end, and allow attachment or detachment of the second cable. The addition of an indicator would have been preferable, otherwise the engineman would have had little warning, particularly during operation in the dark or fog, and instantaneous stoppage was not possible.
Using the lineside telephone, the engineman was informed by the man at the summit when the cable had been attached or detached, and that a train had arrived at the summit or was ready to leave for the quarry. Similar information was passed from the junction siding

trial working on the SDR. There was only a nominal track gauge difference of ½ inch between the two railways, so there was a chance of success. But the loan was unsuccessful, because the locomotive frequently derailed. A house close to the siding on Snailbeach mine was damaged as a result, the evidence still being visible in the 1960s. *Sir Theodore* was soon returned.[12]

Concrete haulage cable return wheel housing on the Granham's Moor quarry tramway in 2012. Rope abrasions show that this housed a cable return wheel under the track. This view looks up the top part of the west side incline to a cutting ending at the summit. There is no evidence of any drum house. Based on this, it is believed that a main and tail cable haulage system was used on the tramway. A winch engine at the quarry, using two cable drums alternately powered or braked, could pull wagons in either direction between the quarry and the cutting shown here. A second operation, using what was the tail cable on the westbound haul, was needed to lower or raise wagons on the west incline below this point. M. J. SHAW

Top view of return wheel housing, with in and out channels for haulage rope left and right. In the centre are a spindle (probably for the return wheel), a piece of steel haulage rope about 7/8 inch diameter, and three dog spikes

The GMQCo planned to build a rope-worked tramway over the Stiperstones ridge to join the SDR. In September 1905, Henry Dennis announced an unconditional agreement with Toye & Co. to carry stone from the junction point for fivepence per ton, subject to a minimum quantity of 15,000 tons per year-ending 30 September. The board agreed to the loading stage at Pontesbury being extended to deal with the new traffic. Dennis also insisted, following the trial of *Sir Theodore*, that '*... one locomotive must be repaired or a new one bought.*' In December 1905, Dennis & Son placed an order for a new locomotive with W. G. Bagnall Ltd of Stafford.[13] The railway relied heavily on the new quarry trade, and mineral traffic increased by a modest 2,000 tons in 1905. In February 1906 the SDRCo board agreed to provide a siding to join the quarry tramway, and to buy the locomotive from Bagnall for £970 by hire purchase. Possible future development was considered. A piece of land was bought in Pontesbury '*... for the purpose of a siding ...*' from Rev E. M. Furley of the Old Rectory, Pontesbury in June 1906.[14] This was the irregular plot on the west side of the railway sidings. For many years it was put to no use other than being let for grazing.

In February 1906 the new locomotive *Dennis* was delivered, the SLMCo granting debentures totalling £1,600 for this and the other capital outlay linked with the Granham's Moor venture.[15] The GMQCo dominated traffic and the mineral tonnage in 1906 trebled to 21,200 tons. The only other major freight was from the SLMCo and the Halvans Co. Other accounts, such as the Minsterley Baryta Co. buying barytes from Snailbeach mine, and F. W. Read who leased the Snailbeach barytes mill, amounted to very little. Typical monthly accounts for Read were £2 6s 10d and 15s 9d, whereas the GMQCo paid £79 12s 4d for the last quarter of 1906. The GMQCo paid quarterly, following settlement by the SCC, creating cash flow problems for the SDRCo.

Henry Dennis died in June 1906. Following this and the death of his successor Francis B. Owen a year later, the common board of directors was Henry Dyke and Walter Pen Dennis, William Edwards and Longueville solicitor Martin Benson Lawford. It stayed that way until the SLMCo dissolved in 1912.[16]

In 1907 mineral traffic again increased to almost 33,000 tons, while train mileage increased by 192 miles and the line still ran at a loss. High locomotive mileage was needed in order to deal with wagons for the quarry. But in 1909 mineral traffic climbed to a record 37,400 tons, and the

Cutting on east side of summit in 2012. This formation involved rock blasting. Paul Teather and Mike Shaw are seen at the cable return wheel housing. At this point, above the return wheel, the wagons of stone were held prior to lowering down to the private siding near the SDR.

Two bearing blocks for the return wheel with nuts and bolts

Looking downhill and westward in the major cutting on the west incline in 2012. Shallow embankments also survive.

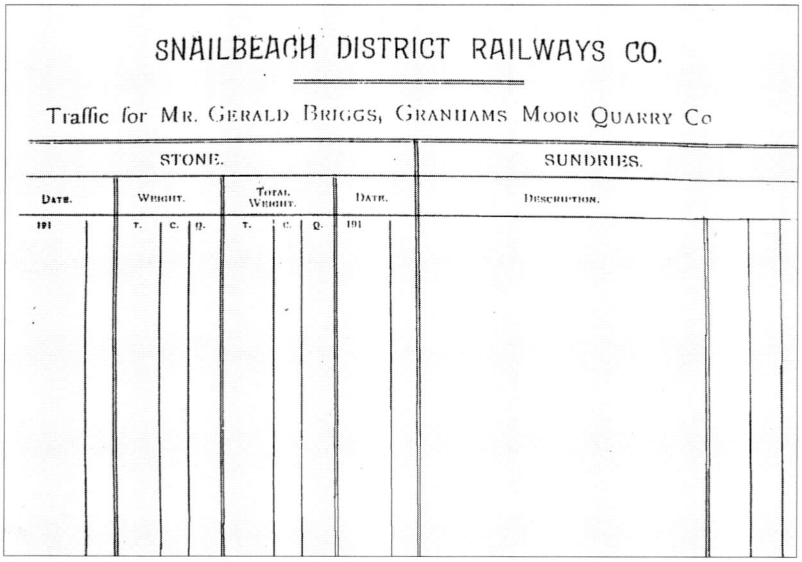

SNAILBEACH DISTRICT RAILWAYS CO.

Traffic for Mr. Gerald Briggs, Granhams Moor Quarry Co

STONE.							SUNDRIES.	
Date.	Weight.			Total Weight.			Date.	Description.
191	T.	C.	Q.	T.	C.	Q.	191	

A blank carriage account form, Snailbeach District Railways Co. to Gerald Briggs, Granham's Moor Quarry Co.

railway made an operating surplus. The railway company, at Lord Bath's insistence, resumed paying his chief rent of £22 a year rather than £5.

In 1908, the SLMCo attempted to negotiate a new lease with the Earl of Tankerville for the Vessons mine, into which the main Snailbeach vein passed from Lord Bath's land.[17] The terms offered by Lord Tankerville included unrealistic royalty rates and the company declined to renew. This led to an agreement in September 1909 with Lord Bath to cease pumping and abandon the mine below the 112 yard level.[18] With the fuller extent of the vein being unavailable, there were insufficient prospects and the mine was run down. The SLMCo liquidated in 1912, and the Halvans Co. took a new lease of Snailbeach mine including the underground as well as the spoil heaps. There was concern as to who would take ownership of the railway. Lord Bath was asked if he would buy the debentures totalling £6,300 for £2,000 and take all the railway shares for a shilling each (£100), and at the same time make no claim for dilapidations of the mine. Lord Bath refused this offer. How ownership was settled has not been established, but Lord Bath took some of the shares.

Late in 1911 the GMQCo asked for a reduction in carriage rates, but was told that the rate was already so low that this was impossible. Then in 1912 mineral traffic slumped again to a little under 9,000 tons, resulting in a loss of £85. The running down of the SLMCo contributed to this, but the Halvans Co. also traded poorly. Mineral production was affected by the shortage of coal caused by the miners' strike in the spring of 1912, and all but a few senior men were laid off. In November the SDRCo was unable to buy coal and oil. The directors decided to close the line, and it stayed closed until April 1913. Dennis & Son no longer had its mine company to serve. The closure could therefore be used to persuade the quarry company (which depended entirely on the railway for transport) to negotiate a new agreement for a remunerative level of freight. The problem was that the railwaymen's wages alone far exceeded revenue from actual quarry traffic. In the year to September 1912, the quarry company had shipped less than the guaranteed annual tonnage, and was invoiced for the difference. It then emerged that Toye intended to sell the business. Company secretary Arthur Bardswell (successor to Martin Lawford) made it clear that new terms were needed with the new proprietors, and even suggested that they could haul for themselves subject to a satisfactory agreement. Meanwhile the Halvans Co. had already guaranteed £250 per annum with a fixed up/down rate of 1s per ton for all traffic.

The buyer of the GMQCo was Gerald Briggs, who was at times a director of the Ceiriog Granite Co. Ltd, and with this purchase Dennis & Son no longer had an interest. Under his ownership the GMQCo then shared the office of the Ceiriog Granite Co. Ltd at Bank Chambers, Swan Hill, Shrewsbury and later at 8 Church Street.[19] The SDR directors' final offer to Briggs was that he would need to guarantee £200-worth of traffic from the quarry per year. To ease the railway company's cash flow problem, the money would be paid monthly pro-rata with quarterly settlement of any difference. Carriage rates were to be fivepence per ton for down traffic and tenpence per ton for up traffic of all types. Briggs agreed, and the railway re-opened on 15 April 1913. Then in September Briggs asked for reduced rates, and suggested that the agreement obliged the railway to transship for him. The SDRCo refuted this and said it would like still higher rates because of continuing loss.

The Great War

During the war, traffic again increased considerably, and relations between the companies settled down. In 1914 the Halvans Co. renegotiated a minimum freight of £200 per year. Its barytes traffic rose from 2,400 tons in 1914 to 6,900 tons in 1916.[20] Granham's Moor stone traffic improved, peaking at 16,000 tons in 1917. The GMQCo had its own wagons which were hired to the Halvans Co. and the SDRCo. These were side-tipping wagons, which were suitable for rail-road transshipment at Snailbeach wharf, and hopper wagons.[21] In the *Contract Journal* of 10 May 1916, the GMQCo advertised its need for 'a small loco 2ft 4in gauge'.[22] Its intended use and whether or not one was bought are unknown. It was needed possibly because of the unreliability of the SDRCo's only locomotive. During the war, general road repair planned by the SCC discontinued, and road stone like other raw materials was commandeered for the war effort.[23]

In 1916 the Shropshire Lead Mines, now owned by Shropshire Mines (1916) Ltd, began using the line. The principal shareholder of Shropshire Mines (1916) Ltd was Sir John F. Ramsden of Bulstrode, Buckinghamshire.[24] By 1917 the railway was dealing with Shropshire Mines direct. The company expected to use twelve wagons regularly and to ship 1,300 tons per month, but in practice the account was far less. This traffic was not to last long, because Shropshire Mines was to build its own aerial ropeway between Minsterley and the Bog mine. Although utterly impractical, a roadside tramway, probably from Snailbeach and to the Bog, was also considered. In 1916 N. L. Corry, an heir of the Listers (landlords of the Bog mine) and W. Bridgeman MP both had a brief correspondence with the Marquis of Bath about the proposed ropeway. Bridgeman hoped that Lord Bath would allow the ropeway to be built across his land, on the grounds that it would keep engines off the roads, thereby reducing the rates, and would increase employment. Another letter Lord Bath received from an unnoted source said that the ropeway was being opposed because of its potential effect on the SDR. Ultimately his Lordship agreed to it being built, and with the SDR struggling to make ends meet, even with the improved traffic, he might have felt no obligation to take sides. Construction of the ropeway began in 1918 and it was operating by 1919.[25]

By 1907 the railway fences were already old, rotten and under continuous repair. Around 1916, company secretary Arthur Bardswell battled with Lord Bath's agents concerning renewal of the lineside fences bordering Lord Bath's land. The fences were now beyond repair, and it was difficult to keep livestock off the line. Bardswell said that to repair the fences at the estimates obtained, the line would have to close. This is a measure of the railway company's poverty, despite the increased traffic. There was a practical problem in that wire makers were all involved in munitions work. Near the end of the correspondence, in the second of two lengthy letters in which he used the title phrase of this chapter, Bardswell wrote:

> ... I certainly hoped [for] more consideration ... No doubt the present Lord has not derived much benefit from the Mine but the late Lord received very large Royalties and I think it is only right to say that His Lordship always met the mine in a most generous manner. The Mine, on which the Railway was really dependent, was ruined by the action of Lord Tankerville asking outrageous terms when his lease expired and if he had granted a fresh Lease on terms which would have enabled the Mine to work, the extreme probability is that it would have been working at the present time, and if so, at the present price of lead, the Lessors would have been getting a good income. I still venture to hope that His Lordship will be able to meet the Railway Company to some extent.

Bardswell then mentions his own salary is only £5 a year and that other officials work for next to nothing, and continues '*... but still I want to do what I can with others to keep the Railway going for the benefit of the District ...*' Despite Bardswell's persistence, Lord Bath declined to provide fence posts or any other help. Eventually some work was done, but the board decided it could afford no more by March 1917. The two miles of new fencing erected in 1916 cost £69 and a further two miles cost £108 in 1917.[26]

The Buxton Hill branch of the Shropshire Mines Ltd aerial ropeway, seen at Burgam Corner, below the transfer station which joined it to the main line between the Bog mine and Malehurst. COURTESY M. J. SHAW

The war was biting in other ways in 1917. The North Wales collieries previously used were not allowed to supply the SDR with steam coal. The Staffordshire, Warwickshire and Shropshire District Coal & Coke Supplies Committee directed the company to the Madeley Wood Colliery Co. The company spent £92 on steam coal in 1914. This escalated to £290 in 1918, and this itself was a doubling of the 1917 cost. For sleepers to refurbish the line, the engineers were referred to the Assistant Controller of Timber Supplies. It is in 1917 that the first mention is found of the railway hiring wagons from the GMQCo. The railway company's rolling stock was deteriorating.

By 1917 the GMQCo had accepted increased rates of sevenpence per ton down from the tramway connection and one shilling up. Further increases were enforced from 1 January 1918. The Halvans Co. and Shropshire Mines Ltd agreed to a 50 percent increase, while timber merchant Jabez Barker of Shrewsbury (see below) paid two shillings per ton. Briggs agreed to pay 1s 6d per ton up and 11d per ton down, but only '*... for one year or for the duration of the war whichever period should be the shorter ...*' To comply with the agreement, Bardswell suggested that Briggs be given a year's notice of a further rate increase immediately. Despite everything, Briggs was anxious to secure the future of the railway for his company. Unreliable locomotive working was a great inconvenience. In February 1918 Briggs made an offer to take control of the SDRCo. The negotiations looked so serious that Bardswell mentioned it to W. K. Arber, the owner of the Halvans Co., but it came to nothing. Bardswell even suggested that Arber could also make a bid for a controlling interest. Not long before, in September 1917, a letter was sent to the Great Western Railway Co. from Lord Bath's office which also concerned a takeover of the SDR. The proponent of this takeover bid is unknown.

The SDRCo's push for carriage rate increases was driven by the wartime escalation in costs, especially wage rises, and of materials, like steam coal and sleepers. On top of these factors, the Ministry of Labour was promoting a 'war wage' to be paid from January 1918. This was such a great concern that correspondence led to a meeting with the Ministry on the subject in October 1918, when the company explained its position. The carriage rates had barely kept ahead of the rapid increases in expenses, mainly caused by wage claims. Wages for the five railwaymen in 1914 totalled £290 and in 1918 were £490. The SDR was not alone in having this problem. Railway carriage rates were regulated by the Board of Trade under Acts of Parliament, but a new approach was to be adopted immediately after the war (see below).

Wartime timber traffic

Around 1914 the U.K. imported roughly 90 percent of its timber, which could not continue with the coming of war. During the war there was a huge need for timber for military use and in mining, and it was felled across the country on a massive scale. The timber trade was poorly organised and in 1915 the Home Grown Timber Committee was formed under the Board of Agriculture to organise supply. During the war and immediately post-war, the responsibility for timber supply switched to the War Office and then the Board of Trade, until finally the Forestry Commission was formed in 1922.[27]

The Stiperstones and thereabouts had a good covering of timber at the beginning of the war, and the authorities took full advantage of it. On the west side of the hill the Snailbeach line was convenient for hauling the timber out. By the end of the war, the entire west flank of the Stiperstones from Callow Hill to Crowsnest Dingle was stripped of useful trees by the timber merchant Jabez Barker.[28] Convenient stockpiling sites were made for loading trunks onto railway flat wagons, like Snailbeach station and a point near the Farm Lane crossing.[29] As a result, goods traffic jumped from £8 in 1916 to £109 (1,115 tons) in 1917, and £125 (1,265 tons) in 1918.

Post-war and the take-over

In February 1919, the locomotive was out of service for some time, and the GMQCo had used horses to haul empty wagons. The GMQCo then claimed there should be no carriage payable for the affected traffic. Bardswell rebuffed this, saying that a brakesman was used for all down traffic, and there should be payment for use of the railway. In April the disputed charge was still due, but on learning that the quarry company had paid much more for horses, labour and storage than it owed, Bardswell suggested that the claim be dropped. Although the rates chargeable under the Order Confirmation Act 1892 were already greatly exceeded, in June 1919 it was proposed to the quarry company that the rates be increased to 1s 6d per ton up and down for all traffic from 1 July. The GMQCo refused to pay it, yet continued to settle at the old rates. Following a meeting, the GMQCo was told that the railway could not continue at the old rates, and was given notice to terminate the agreement on 30 September 1920. After a further meeting the rates were agreed at 1s 3d down and 1s 6d up from 1 September 1919. Compromise was necessary: the quarry company needed the railway, but the railway company could not provide reliable haulage for the traffic.

With the end of the war, much had to be done to stimulate private trade. Some relief for the Board of Trade and more freedom for the regulation of carriers came when the Ministry of Transport was created by an Act of Parliament in August 1919.[30] The new Minister acted quickly to ensure that transport revenue could be increased. A Rates Advisory Committee was formed to investigate and recommend new national rates for freight and passenger transport. It recommended in December 1919 that railway carriage rates for all Class A goods and minerals including stone and spar be increased by 30 percent plus a flat rate threepence per ton. On learning the news, Bardswell wrote to the GMQCo that *'My Directors find that they have to increase their charges in accordance with the terms of the Ministry of Transport Act 1919.'* The quarry company was charged 1s 10½d per ton from 15 January 1920.[31]

Just over the border in Montgomeryshire before the Great War, the Criggion road stone quarry of the Pyx Granite Co. (PGCo) began to thrive with the opening of the standard gauge Shropshire & Montgomeryshire Light Railway (S&MLR). This line was one of the minor railway group controlled by H. F. Stephens, and hauled to Shrewsbury via a branch line all the stone the quarry produced. In 1912 the S&MLR carried over 16,000 tons of Criggion stone.[32] Gerald Briggs of the GMQCo soon took an interest in the Criggion operation. From 1914 the GMQCo became part lessee (with the Ceiriog Granite Co) of the PGCo's Criggion quarry lease. In 1916 the GMQCo became its sole lessee, and took out another lease in its name alone of about 2½ acres of land at Criggion on which to erect buildings. From January 1916, it also leased 42 acres for quarrying not far from Criggion, at Pritchard's Hill near Crewe Green. Criggion output improved during the war, growing to over 45,000 tons in 1919. In November 1919 Gerald Briggs bought outright the Criggion quarry business from the PGCo. By 1920 Criggion was producing at least three times the wartime peak output of Granham's Moor quarry.

At the end of March 1920 the SDR's locomotive *Dennis* was withdrawn, but despite this there was occasional traffic using horse haulage. In May, Gerald Briggs mentioned to Bardswell that the S&MLR was interested in buying the SDR. Inquiries from H. F. Stephens quickly followed, in which he asked if the railway company had any land between Pontesbury Junction and Pontesbury station. It was intended to extend the line and introduce a passenger service, and in anticipation Stephens had multi-line free travel passes made which included the SDR.[33] Like every other extension plan, it never happened, and neither did a passenger service. Contrary to earlier accounts, including a report by *The Locomotive* magazine of 15 October 1926, Stephens did not buy land at Pontesbury for the building of an extension. The company reported owning only one acre of land (more accurately 1 acre, 1 rood and 2 perches) not part of the railway or stations both before and after Stephens took control. This was the plot bought by Dennis & Son for a siding at Pontesbury.

Bardswell offered a price of £3,965 to Stephens to buy a controlling interest in the company. Outright ownership was not legally possible because some shares were missing. The debentures were valued at £3,150 and the shares at £500. Compensation to the company's officers would be £105 (or 100 guineas) for Dennis & Son and the same for the secretary. The balance was for legal expenses. There was also a similar offer made to Brig Gen R. S. Stronach of Roadrail Loco Tractors Ltd, 1 Dover Street, London W.

In June 1920 the railway had not worked officially for many weeks, and the railwaymen had been laid off. H. F. Stephens met the railway company directors on 31 July 1920 and the earlier offer was confirmed in writing. The offer expired, but Stephens wanted a lower price, and in October he wrote about the increased need for permanent way repairs since the withdrawal of the locomotive. This was because of the use of horses.

In December 1920 the GMQCo owed the SDRCo £337 10s 9d and Bardswell threatened to stop their traffic. This threat was difficult to put into practice. Bardswell repeated the threat and considered legal action. But by February 1921, the quarry company had negotiated a reduced rate of a shilling per ton up and down backdated

This photograph has been dated at around 1905 by the late Ken Lock. Plentiful timber in the carpenter's yard and the steam emanating from various chimneys tells us that the mine was quite busy. At this time the SLM Co was exploring eastwards under the lease of Vessons mine from Earl Tankerville. Some short time afterwards, the company declined to renew this lease owing to his Lordship's demand for unrealistic royalties. Exploration ceased, the mine was flooded to the 112 Yard Level stopping deep mining, and the company liquidated in 1912.
Steam is being raised in the locomotive shed. There is a line of hopper wagons and perhaps a pair of flat wagons on the long siding, and a low sided wagon on the carpenter's yard siding. This is the only known view showing the coal yard on its masonry revetted platform, bringing coal in above the central boiler

house which was new in 1881. Several coal wagons and what was probably a small office can be seen on that level.
The tramways exiting Landing Level can be seen, and the waste tramway to the black hillock which passes the smiths' shops. This starts at a tip from one of the Landing Level branches, as it did in the 1870s. The other branch now runs on a curved trestle to the crusher house, and not as in 1885. The chimney visible next to the plume of steam is that of the old boiler house, which has yet to be truncated. While most dressing machinery is under cover, the old buddles near the locomotive shed have been replaced in the same outdoor area by a large and undetermined piece of machinery. The raised cistern nearby has gone and its revetted hillock is mostly broken down. COURTESY MRS. PEGGY CHIDLEY

to 26 March 1920 (less than the pre-Ministry rates). As on earlier occasions, the GMQCo had arranged its own haulage to replace the withdrawn locomotive since that date. The locomotive *Dennis* did not re-enter service. In his February 1921 letter Bardswell wrote to the quarry company that '... *my Directors have not yet succeeded in obtaining such a tender for a locomotive as they can see their way to accept.*' Although operating, the SDR was becoming utterly run down, and it was in 1921 that the last Granham's Moor stone was carried. With the S&MLR providing Briggs with a good steam-hauled service for Criggion quarry and having the haulage capacity to support increasing output, there was no obstacle to the closure of Granham's Moor quarry. From this time in 1921, the SDRCo had only the minimal traffic from Snailbeach left, and were waiting for Stephens to buy the line. They reduced every operating and maintenance cost possible. For the SDRCo's 1921 Annual Report, Dyke Dennis refused either to sign a safety certificate for the track and rolling stock or even to be implicated. The Ministry of Transport reported in 1921 that the line had recently closed.[34] However, there is nothing specific about this report, and it might refer to the recorded temporary closure of 1920. Traffic on the railway continued and revenue was received through 1921 and 1922, until Stephens completed his purchase.

The expenditure recorded in the Annual Reports shows how this was done. In 1920, until withdrawal of the locomotive, £160 was spent on running costs including the driver's wages. In 1921, nothing was spent on locomotive running or repairs and only £8 on wagon repairs. Only wages for the transshipper (the junction man) were paid, of £17 as opposed to £29 in 1920. In 1922, only the fences were maintained, at a cost of £50, and again only wages were paid for the 'transshipper' of £84, but such an amount probably includes other costs: horse working. The Annual Report for 1922 also records in the Engine Mileage section that '*The Company had no engine working during 1922, and the traffic was worked by Horse.*' In fact this had been the case since early 1920. R. T. ('Tom') Oldfield, a son of mine manager William Oldfield, is one haulier known to have done this work.[35]

In the takeover negotiations both sides held out from mid-1920 to 1921 – certainly nothing was recorded by Bardswell – then in February 1922 a possible increased offer was suggested by Stephens' solicitor W. K. Minshall. This would only be on receiving assurance that there had been no serious drop in traffic between 1920 and 1921. Bardswell admitted that traffic had dropped by about half owing to the engine breakdown (which caused, in Bardswell's words, '*the inevitable loss of the stone traffic*'). The S&MLR management might have known this already through contact with Gerald Briggs, and it would have been a useful bargaining chip to keep the price down. In December 1922 the sale was concluded, with the shares being transferred to Stephens, and indemnity arranged for the 40 missing shares. The first replacement locomotive was also bought.[36] The *Newport & Market Drayton Advertiser* of 9 December 1922 announced in a news item the intention to re-open the line for general goods traffic and hopefully for passenger traffic later.[37] Bardswell completed the 1922 accounts for a final salary of £10, and all company documents were handed over in April 1923. The line did not re-open with steam traction for some months after the purchase.

The trade from the Granham's Moor quarry was, in its first flush, the saving of the railway, but it was only the unsuitability of the local lanes to heavy traffic which brought it about. The Granham's Moor trade then slumped before the war. If traffic could have continued at 1908 to 1909 levels without the Great War, and if the Tankerville lease could have been renewed at an economic price, then it might have been possible for the railway to recover under Dennis & Son. The war increased the demand for stone and minerals, but it also brought inflation in the price of labour, fuel and materials, and the railway company was running to stand still – simply 'for the benefit of the District'. Traffic from Granham's Moor quarry and Snailbeach mine was never enough to cover all the costs. There was no money for capital investment (for example to replace the locomotive) and much of the revenue slipped away after the war, when Gerald Briggs found better opportunities elsewhere. Only the traffic from Snailbeach remained then, and Dennis & Son saw no future for the line. They must have been quietly surprised that H. F. Stephens did.

CHAPTER EIGHT

New prosperity and the end of steam 1923–1946

With the sale of the company, its management shifted from one principally interested in mining to one which specialised in minor railways. H. F. Stephens ran his railway engineering and consultancy practice from offices at 23 Salford Terrace, Quarry Hill Road, Tonbridge, Kent, and this became the Registered Office of the railway company. The premises consisted of a basement and three floors. The staff of about 17 in the 1920s included Stephens' 'indoor assistant' or principal clerk J. A. (Arthur) Iggulden, and two other assistants. There was a general drawing office and civil engineering department on the first floor, headed by 'outdoor assistant' W. H. (Bill) Austen. Engineering and operating aspects of the railway were managed by memoranda sent by post from Tonbridge. Other business was handled by the Abbey Station, Shrewsbury office of the Shropshire & Montgomeryshire Light Railway, with the two lines treated as a unit. There was at least an annual inspection of the railway.[1]

The SDR was Stephens' personal property. At the first meeting under the new ownership in February 1923, Stephens, H. Montague Bates and J. L. White were elected directors. The former auditor G. M. Jenkins was re-elected. Stephens was also the chairman and R. F. Bowmaker was secretary and accountant.[2] The railway company's new management inherited very little traffic except that of the Halvans Co. Ltd, which alternately carted the minerals and coal or used the railway according to cost.[3] Survival, and profit, would depend on finding new users.

It also depended on having sufficient serviceable motive power and wagons. The Kerr, Stuart locomotive *No.2* was delivered to the line on 10 March 1923. This was the earliest that a steam-hauled service could have resumed. *Dennis*, now also known as *No.1*, was still out of service with an uncertain future, and a further two replacement locomotives, originally built in the USA by the Baldwin Locomotive Co. for the Ministry of Munitions, were not delivered until April 1923. Fitter Mr J. Webb was still working on modifications to the frames and brake gear of these two locomotives in mid-May. Stephens himself visited the line in May, and reported by memo on 9 May to Bill Austen on progress with wagon repairs. There was plenty of work to do to put the railway in running order, but little money was spent on the track in 1923.

23 Salford Terrace, Tonbridge, the former office of H. F. Stephens' minor railway empire, pictured in 2006. A. AUSTIN

MINOR TRADERS AT SNAILBEACH AND THE EFFECT OF ROAD TRANSPORT

At Snailbeach mine, the Halvans Co. continued to maintain the primary lease of both the underground workings and the main spoil heaps. Then in about 1923, Edgar Charles Gray, a mine manager for Shropshire Mines Ltd, took the plot formerly occupied by F. W. Read next to the railway line at Snailbeach. Here he set up a spar gravel works, re-working the old Black Hillock (spoil heap) on the opposite side of the road.[4]

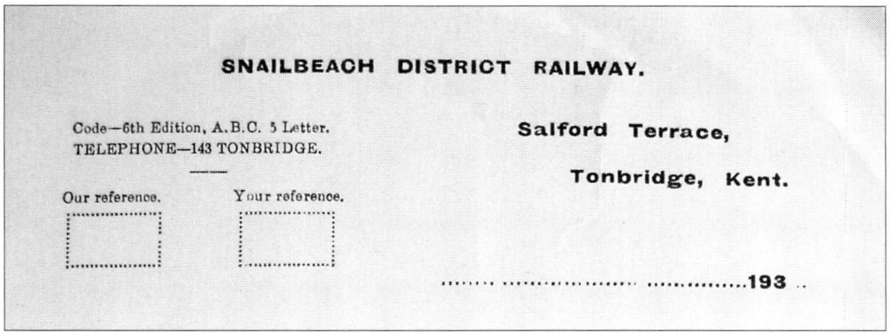

SNAILBEACH DISTRICT RAILWAY.

Code—6th Edition, A.B.C. 5 Letter.
TELEPHONE—143 TONBRIDGE.

Salford Terrace,
Tonbridge, Kent.

Our reference. Your reference.

..................193 ..

Letter-headed cream wove notepaper of the 1930s. COLONEL STEPHENS RAILWAY MUSEUM ARCHIVE

No.2 with David Jones driving in 1936. The original Roscoe cylinder lubricator has been removed, probably at the same time as the large patch was fitted to the side of the smokebox, which is still the original. Note the lamp bracket still in place. But the smoke box door is new, and has hinges which are much closer together. In all views of this, the hinge pin always projects above the upper hinge.

Gray used the railway, with his wagons left on the main line for loading, but later a siding was added.[5] He moved from the area in the late 1920s and his Snailbeach works ceased.[6] This was a very small business, and appears not to have been very steady in its output. In a seventeen week period between September and December 1927, Gray used an average of four or five wagons per week, with the most usual numbers being from three to six. For three of the weeks only one or two wagons were needed, and in only two of the weeks were eight and ten wagons required. At say three tons per wagon, one can easily estimate the small amount of traffic from Gray's works.

Although there had been a drift of population from south west Shropshire to the towns, longstanding mining families worked hard to make a living at Snailbeach rather than move away. Occasionally they traded to bring in extra cash. There was a small trade in coal on the railway by a Mr Henry in the mid-1920s, but almost nothing is known of it, and Henry has not been traced. More is known of the coal trade of Clarence Rowson, a Snailbeach resident. For some years until his death in 1932, he rented a shed (more properly perhaps, an open-topped store) in the Snailbeach wharf yard for this business.[7] Wagons loaded with his coal were left on the wharf siding next to the yard, sheeted over and protected with barbed wire.[8] Clarence Rowson's death contributed to reduction of coal freight from 246 tons in 1931 to 140 tons in 1932, and 34 tons in 1933. He was buying and selling a few tons a week. In the late 1930s the store was fitted with a roof, and rented by the Shropshire County Council as a garage for the District Nurse Charlotte Hand's car.[9]

By 1929, the Halvans Co. had been bought by Charles Moore & Co. of Lymm, a chemist and salt manufacturer using barytes. Charles Moore & Co. had bought Snailbeach barytes for some years and had no interest in the spoil. This left an opportunity for a sub-lease of the main spoil heaps. Moore continued mining barytes in Perkin's Level and Lord's Hill Mine in the Upper Works at Snailbeach very briefly,

Spar works in the white hillocks, late 1920s. In the foreground is a dolly tub, a final stage in the ore washing process.
COURTESY K. C. LOCK

Snailbeach station in the late 1930s. Clarence Rowson's old store is having a roof put on it to shelter Nurse Hand's car. Hughes's bridge has had its stone parapets replaced by iron railings to improve visibility for motorists at the double bend. H. F. G. DALSTON

using Joseph Roberts (the Snailbeach Barytes Co.) as a local contractor.[10] Barytes traffic on the railway ceased in 1930.[11] Then some barytes was raised sporadically in the 1930s but was removed by lorry.[12]

The Gravel Trading Co. Ltd was registered in 1928. Taking a sub-lease of the spoil heaps, it shipped spar gravel products via the railway through the 1930s.[13] The former railway company director William Edwards traded as the Gravel Trading Co. Ltd and the Whitstone Company from Ruabon.[14] By 1930 over a dozen customers of the Gravel Trading Co. were being sent consignments of spar products in various sizes and grades, amounting to some two dozen wagon loads a week. Wagon bookings in Edwards' name show that he used the railway for carrying minerals separately from the Gravel Trading Co. from as early as 1927, and until as late as 1934. No other companies at Snailbeach being known of at this time, it is presumed that these were for his Whitstone Company operation. In late 1927 his wagon requirements averaged twenty a week, ranging from nine to 31. Some figures for 1930 show that this had decreased to ten or less (although this was only over a two week period), and that Whitstone specialised in white dust and gravel products. In the 1930s Edwards' two companies were the only mineral users of the railway at Snailbeach. Late in the decade the Gravel Trading Co. alone remained, and by 1939 was making enquiries about road haulage of spar to Liverpool.[15] The company last used the line in 1940.

It was not only mineral traffic being taken away from the SDR by competitive road haulage, but also general goods, although the line carried very little. The directors Austen and Ramsay saw for themselves in 1939 that the local Great Western Railway lorry was delivering goods and parcels to Snailbeach, and according to their informants, almost daily. The men were encouraged to try and get the traffic onto their

Gravel Trading Co. rail delivery post card. COURTESY SHROPSHIRE CAVING & MINING CLUB

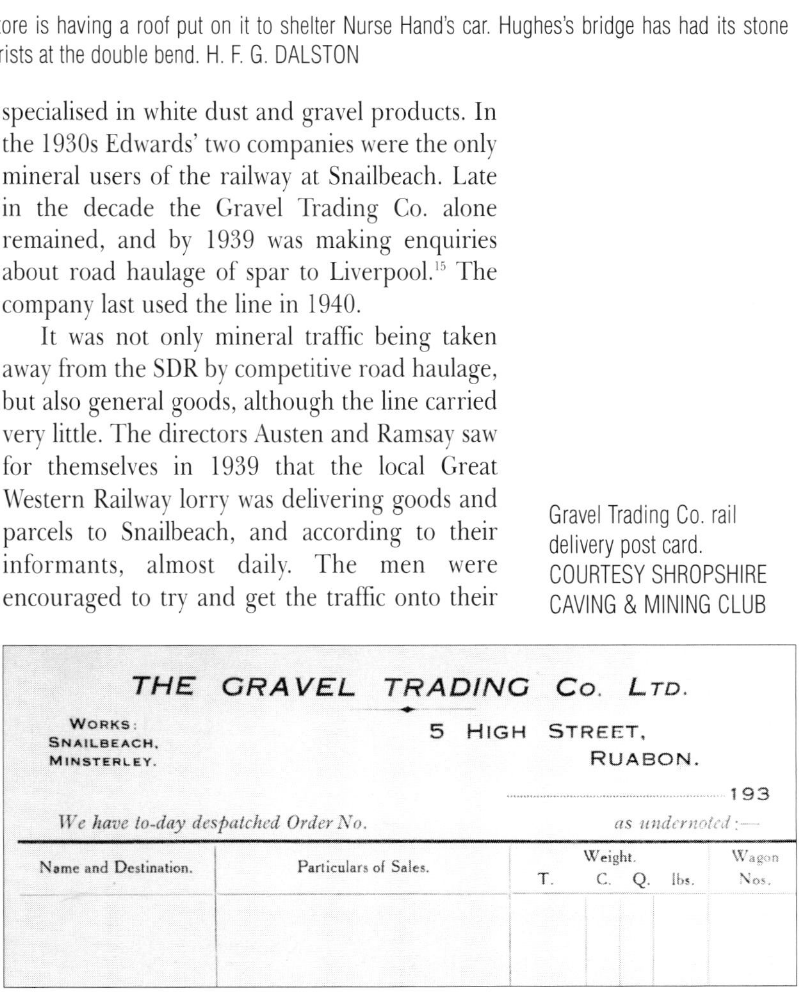

THE GRAVEL TRADING Co. LTD.

WORKS:
SNAILBEACH,
MINSTERLEY.

5 HIGH STREET,
RUABON.

193

We have to-day despatched Order No. as undernoted:—

Name and Destination.	Particulars of Sales.	Weight. T.	C.	Q.	lbs.	Wagon Nos.

Looking north, Gray's spar gravel works is shown in the late 1920s. This works occupied the same plot of land as the earlier barytes mill built by Haywood. Gravity runs were regularly made from here. Originally there was no siding and wagons were loaded as they stood on the line. The wagon nearest the camera is one of the screw braked wagons with side and end doors by W. G. Bagnall. In the background is the Halvans bridge, with the main spoil banks showing pale grey and Snailbeach Coppice behind them. The dark heaps nearer the camera are the poorer spoil from the Black Hillock worked over by Gray, stacked up on the near side of the road. A short tramway to bring spoil across the road was approved by Shropshire County Council but whether it was built is unconfirmed. COURTESY K. C. LOCK

line, but to no avail. In the 1920s general goods (possibly including coal) carried by the railway declined from about 500 tons to under 100 tons a year.[16] From 1931 to 1947, carriage of coal, coke and patent fuels was recorded separately from either goods or minerals. In 1931 coal amounted to 246 tons of a combined 250 tons, while in 1938 goods carried was 109 out of only 112 tons combined. By the end of the Second World War goods and coal traffic together had shrunk to around 20 tons.

Hayward's Quarries Ltd and Callow Hill Quarry

The railway company will have been encouraged by the new traffic in 1923, but it was still very little. The potential for substantial new traffic lay close by the railway in the form of fresh deposits of stone suitable for road metal. This was soon to become a reality, and the railway was unlikely to have survived without it. There is tangible evidence that Stephens investigated tapping a quarry south of the existing railway (see Appendix 4). It would have involved building an extension to the railway, but fortunately this proved unnecessary.

In the 1920s, S. J. Hayward of Montford Bridge became either the owner or lessee of large amounts of land on the Stiperstones. This included Callow Hill, among other land acquired locally when the Marquis of Bath and Earl Tankerville sold their estates. On 24 March 1926 Thomas Hutton made a lease to Hayward for a stone quarry on Callow Hill which specifically excluded any metal ore, but which permitted a siding joining the SDR.[17] In time Hayward bought the freehold.[18] As well as stone, there was valuable timber and game. The timber merchant Jabez Barker had previously farmed the timber (see Chapter Seven) and was the first to take a significant interest. On 2 January 1923 the Marquis of Bath sold the woodlands known as Hope Coppice, Maddox Coppice, Snailbeach Coppice, Resting Hill and Poles Coppice to Jabez Barker Ltd. On 10 July 1925 Barker then bought

Pontesbury sidings, 1932. The line of hoppers on the exchange pier might carry spar products from the Gravel Trading Co., the Whitstone Co. or indeed stone from Callow Hill quarry. The Shropshire County Council had begun using lorry transport to take road stone from Pontesbury, but was also using ex-Hayward's Quarries Ltd wagons on the branch line railway. A. W. CROUGHTON

Callow Hill quarry scene, late 1920s. Notice the light, portable steel track used at the quarry face. The permanent track to the incline and processing plant is seen in the foreground. BRITISH GEOLOGICAL SURVEY

146 acres of Eastridge Wood and Lordshill from the Earl Tankerville. Excluded was the land leased to the Granham's Moor Quarry Co. Ltd. Then in March 1927 Jabez Barker sold this former Tankerville property to Hayward's Quarries Ltd, followed on 11 May by a total of 481 acres in Maddox Coppice, Lordshill, Resting Hill and Crow's Nest.

As early as 1922, Hayward was trading as '*S.J. Hayward & Co., Steam Haulage Contractors*' from Talbot Chambers, Market Street, Shrewsbury. Hayward's Quarries Ltd (HQL) was registered in 1925, and shared the same address.[19] Traffic on the line began to improve with Hayward's business. HQL immediately began quarrying road stone at Callow Hill, and stone traffic on the railway is recorded as restarting in 1924, although the quarry lease was formalised a little later. The quarry had no suitable road access for lorries, and depended on the railway for transport. A passing loop was laid by the SDRCo close to Callow Hill quarry. In 1927, Hayward planned major investment. Under an agreement with the railway company dated 6 July 1927, a private siding served at each end by crossovers from the line was installed, and the earlier passing loop was removed. Stone crushing and screening plant driven by a steam engine was erected beside the siding, fed by small tipping railway wagons sent from the quarry on a rope worked incline. Railway wagons parked under the plant were loaded from above.[20] The planned work was in place before 1930.[21] The plant was supplied by Samuel Pegg & Son of Alexander Street, Leicester, which used the trade name Pegson Plant.[22] HQL also operated a small quarry in Crowsnest Dingle until about 1927, delivering stone onto railway wagons at Snailbeach. The stone was shipped from Pontesbury by the branch line, for which the quarry company

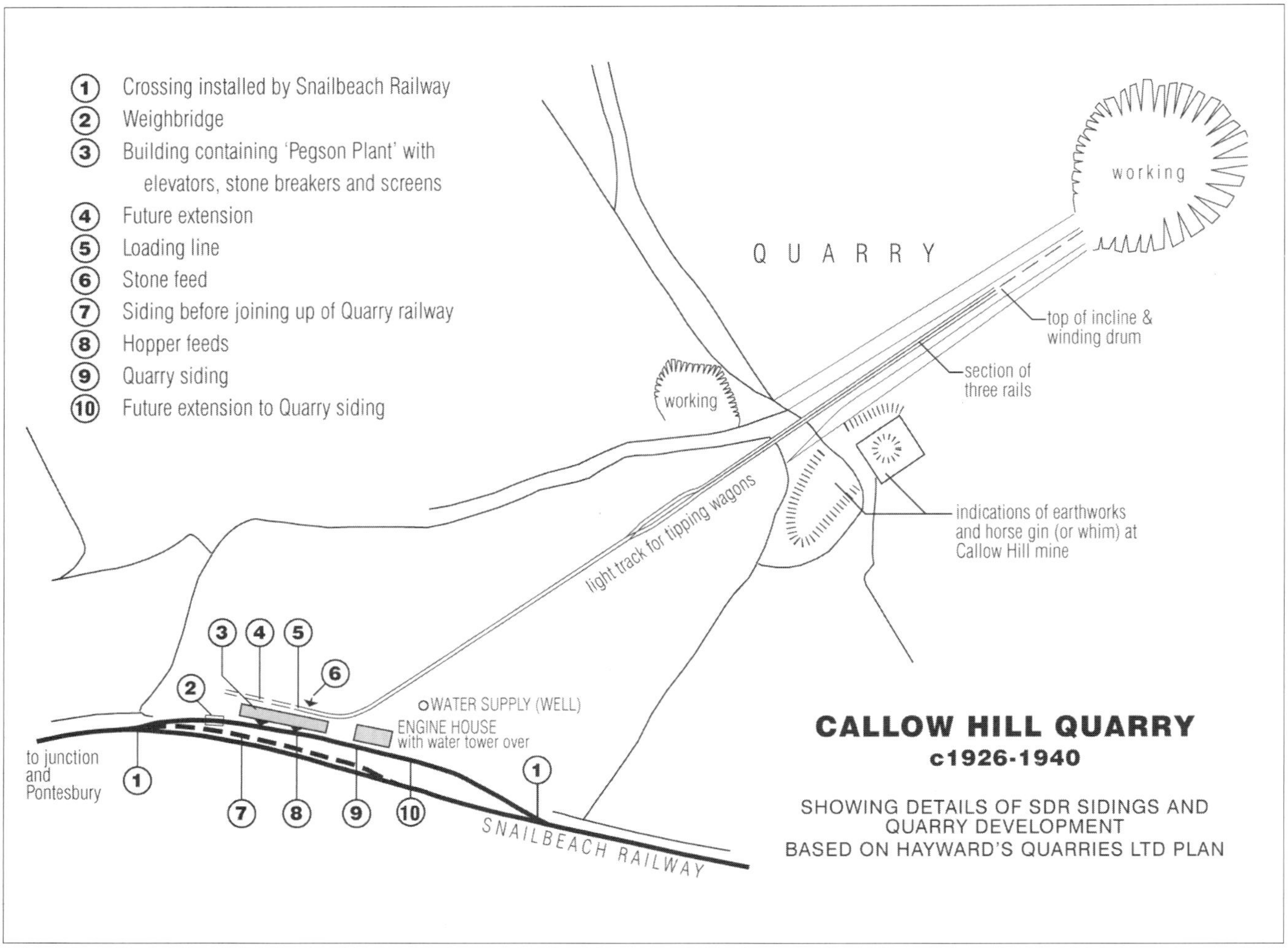

bought nine standard gauge railway wagons.[23] Timber was also harvested from Hayward's land using the railways as transport. Some of the loading of timber was done by removing railway fencing at convenient points. In 1927, Callow Hill quarry employed six men inside the quarry and eight outside, while the other quarries being operated by the quarry company in that year were Nill's Hill (Pontesbury) employing 15 inside and 21 outside, and Upper Mill employing five inside.[24] By the 1940s the office of HQL had moved to Old Bank Buildings, Bellstone, Shrewsbury.[25]

The Shropshire County Council (SCC) was the quarry company's principal buyer of road stone. Making a further agreement with the railway company in 1930, HQL now rented land at Pontesbury for use as a road vehicle yard for the collection of road stone. Hayward then gave the SCC a 14-year lease dated 7 Oct 1931 of Callow Hill and other quarries. The SCC acquired the quarries to give itself control of the cost of road metal in the face of commercial quarry companies. The stock, plant and equipment were purchased outright. This included some plant at 'Minsterley Junction' (meaning Pontesbury Junction) and the standard gauge railway wagons. At the same time, the agreements made between the SDRCo and HQL in 1927 and 1930 were transferred to the SCC. Between 1930 and 1933, HQL and the SCC were loading stone onto lorries in addition to the standard gauge wagons. This was done by using SDR box wagons, and shovelling the stone from them down the embankments north and south of the road. In 1932 the steam engine driving the Callow Hill crushing and screening plant was replaced by a water-cooled Ruston Bucyrus oil engine.

In December 1932 the SDRCo offered the railway undertaking for sale to the SCC. The SCC still saw the railway as the only possible transport out of the quarry, but the proposal presented problems. The railway was legally a common carrier and there was still some use of the line from Snailbeach. County Councils had

Crushing and screening plant and the line at Callow Hill quarry, 26 July 1952. The private siding is on the left, while on the right is the line to Snailbeach. The sidings have been modified since the end of steam traction. The original crossover has been removed, and the remaining check rail on the main line can be made out.
F. W. SHUTTLEWORTH

no powers to operate railways as common carriers, meaning that the SCC would need an Act of Parliament. A benefit was the potential for considerable savings in carriage costs, but this was limited by the shortness of the quarry lease. So the SCC suggested that it would be better for the quarry company to buy the railway and for the SCC to come to an agreement with them. In January 1933 it was learnt that Hayward was prepared to buy the railway company, if he could

The Callow Hill crushing and screening plant close up, looking towards Snailbeach. Note the liberal use of unsawn tree trunks as structural members. This plant was essentially the same as installed by Samuel Pegg & Son in the 1920s, although now driven by a diesel engine. 26 July 1952.
F. W. SHUTTLEWORTH

Rear of a rake of wagons under the crushing and screening plant, looking towards Pontesbury in July 1952.
F. W. SHUTTLEWORTH

get a 'reasonable price' of about £3,500. The SCC therefore refused the SDRCo's offer, and told Hayward that it would enter into a traffic agreement at suitable rates. In the end, no sale was agreed. Hayward could not accept the railway company's asking price of £4,250.[26]

Pontesbury sidings in 1933, showing the as-yet grassy yard, without the lorry-loading bay which was to follow. There is also no tarring plant, and box wagons are in use, which were unloaded down the 'chutes'. Will Jones' office and a sign board (wording unknown) are also seen. The left-hand door undoubtedly conceals an earth closet. Gordon Cooke remembered: 'Ah ['yes' in Shropshire dialect], that were Will Jones' office, and in there he had one o' them old fashioned wind-up telephones.' After a visit in 1926, owner H. F. Stephens sent Jones the following memo: 'Your telephone wire is touching the branches close to the office. You must get the branches cut otherwise it will stop you speaking.' W. E. HAYWARD

The shovelling from box wagons continued in 1933 and afterwards. In 1935, the SCC decided that rail distribution was uneconomical. The standard gauge wagons bought with the rest of the HQL plant incurred siding charges and needed repair, and it was decided to sell them. Also about that time, a tarring plant and a weighbridge were sited in Pontesbury yard and the tarred road metal was also collected by lorry. To help the SCC move away from the laborious shovelling, the SDRCo bought extra hopper wagons previously owned by the Granham's Moor Quarry Co. Ltd. from the British Quarry Co. Ltd. Hopper wagons were used to drop stone chips onto the elevator which fed the tarring plant. The tarring plant siding was extended southwards before September 1938, and a lorry loading bay built beneath it.[27] With these developments, all useable stone was again carried by hopper wagon.

The company's business in the 1920s and 1930s

The period 1923 to 1930 was one of consolidation for the railway company. The barest minimum was spent on the permanent way and wagons in 1923, yet a net revenue of £16 was scraped from 3,500 tons of traffic. There were a couple of years of small losses, but with the new traffic from Callow Hill, tonnages rose from 5,000 to 11,000 by 1930, and profits of £100-plus were seen. The old regime sometimes turned in a worse profit for much more traffic. One reason for the new success was the reduction of railway staff to a minimum. The Dennis regime employed five men in its last active period. Stephens reduced the locomotive crew to one and did not employ an extra labourer. Instead the three men could usually handle the work, which was sufficiently light, by helping each other (see Chapter 15). A few weeks' temporary use of an assistant platelayer is recorded in 1933. Other skilled staff could be seconded from Kinnerley depot on the Shropshire & Montgomeryshire Railway. Dennis & Son had no such support, and had to pay contractors for specialised work.[28]

In October 1931 H. F. Stephens died, and the railway shares and debentures were inherited by his favoured assistants Arthur Iggulden and Bill Austen. Between 1931 and 1935, stone traffic from Callow Hill more than doubled to around 24,000 tons per year. Costs were under control, allowing profits to increase

from £131 to £488 in the same period. In 1937, traffic was almost as good at over 21,000 tons, but unusually high maintenance costs on permanent way and wagons led to a loss of £1,098. In 1938, pre-war traffic peaked at 26,000 tons, and costs had settled down again to allow a profit of £567. In these more prosperous times, payments were made for expenses and special allowances for directors and Tonbridge office employees. The money owed to the debenture holders was also dealt with. In 1935 this debt was reduced to £3,235, and further payments were made in subsequent years as funds allowed, until by the early 1940s only £1,800 was owed. It was also possible to consider modest capital investment in the 1930s. In 1936, it was proposed to install an internal telephone system on the railway (telephones at Pontesbury and Callow Hill on the General Post Office system had been in use for some years[29]). This was postponed, and then dropped altogether when in 1937 it was proposed to erect a new locomotive shed at Pontesbury, which would have reduced locomotive mileage. In the end neither scheme materialised. Instead, the Halvans Co. granted a new lease of the locomotive shed, coal yard and carpenter's shop for ten years from 25 March 1937.

The Second World War and the end of steam

During the war, road stone production dropped to 1920s levels. The carriage rates were increased by ten percent in 1940, and this was the last year that the Gravel Trading Co. used the line. Stone from Callow Hill quarry then became practically the sole source of revenue. Apart from the small and dwindling goods traffic noted earlier, between two and five tons of coal and possibly other fuels were carried per year. This was only a small fraction of the coal then bought by the railway, principally as locomotive fuel, which was delivered in seven to ten ton lots about five times a year. The general decline in stone traffic was noted by the board in 1943, and enquiries were made with the SCC to see if any increase could be made. A net loss of £57 for the year meant that directors' fees and office expenses could not be paid. But in 1944, a profit of £130 was made on the carriage of 8,300 tons of stone. Profits were affected by wartime wage inflation. Wages for the

Pontesbury c.1936, showing the tarring plant and the wagon shelter. On the tarring plant siding, as well as hopper wagons, there is a flat wagon closest to the photographer. On the east siding is an old coal wagon. The end of the disconnected narrow gauge coal siding, which is grassing over, can just be seen beyond and to the left of the coal wagon.
H. F. G. DALSTON

Snailbeach mine from above the Count House, probably in 1936, dominated by the expanse of the white hillocks. Just north of the Halvans bridge, the main line to Pontesbury is hidden by the white hillocks, but it comes into view again at the felspar siding, which has heaps of glistening white felspar beside it. Now only the Gravel Trading Co. Ltd and the Whitstone Co. of William Edwards and supervised by Joseph Roberts were using the SDR from Snailbeach, loading their wagons on the felspar siding. A little further along, the line crosses the Farm Lane, then disappears behind the trees and hedgerows.
From the winding gear marking George's Shaft, there is a 2ft 4in gauge line built on a trestle. In the Halvans Co. Ltd era of underground working, post-1912, barytes was worked from above the 112 Yard Level. The tramway took barytes in mine tubs from rails on the winding cage floor, probably to be dropped either onto a chute or direct to a dressing floor. This work had finished by about 1930. There is a louvred ventilator on the roof of the miners' dry, which has not been reinstated in the recent restoration. Also noteworthy are the derelict buildings and horizontal engine house on the north side of Clink. (Clink is the name of the road between the spoil heaps leading up to the mine). H. F. G. DALSTON

platelayer alone rose from about £120 in 1941 to £135 in 1944 and £180 in 1946.

To counter this slight upturn in fortune, trouble was brewing at Snailbeach mine. At the end of March 1944 the Halvans Co. ceased trading, and its local manager Joseph Roberts who traded as the Snailbeach Barytes Co. took the lease of the mine. This affected the railway company in that it held a sub-tenancy, permitting the use of the locomotive shed and water supply, the coal yard and the carpenter's shop. In June 1944, Austen met Roberts at Snailbeach, when Roberts claimed that '... *certain land on which the Company's rails were laid had been conveyed to him by the Longleat Estate Company* ...' For the company to continue its tenancy under him, Roberts requested a rent of £20 a year. Negotiations were pursued, but matters came to a head when in March 1945 it was reported that Roberts had not only cut off the railway's water supply, but also removed a piece of track on the main line by the wharf bridge, thus cutting off access to the locomotive shed. But because the driver was in the habit of leaving the locomotive on the line below this point, the railway could still operate. The company's solicitors in Shrewsbury, Minshall Pugh & Co., obtained an injunction against Roberts, and a court hearing in August led to a settlement. Roberts agreed that the SDRCo owned the main line from Pontesbury to Snailbeach and the branch line leading to the locomotive shed. A 14 year lease from 29 September 1945 would be given of the shed, a short siding leading off the branch line to the shed and the water supply, for a rent of £10 a year. He also agreed to pay £50 towards the company's costs, and in return the company agreed to withdraw its counterclaim of £87 for loss of water supply and the removal of the length of track.

In 1945 and 1946, the last two years of steam operation, the tonnage began to climb again, but locomotive costs were excessive and losses were heavy. A runaway of three loaded wagons of stone on 3 November 1945 was far from helpful. The incident also involved another six wagons, with all nine of them derailing and suffering substantial damage. It resulted in Bill Austen chiding Will Jones about the cost to the company of having nine wagons out of service and having no money to replace them. In about January 1946, the piece of grazing land north of the road vehicle yard was sold to a Mr Lakelin. After the resignation of Dickie Preece in November 1945 the locomotives were driven and maintained by George Edwards, a fitter employed by the SCC (See Chapter 15). Early in 1946 the Board considered the possibility of closure, and awaited a response from the SCC *'... as to whether they could assist in any way to keep the line open for traffic ...'*

As instructed by Bill Austen, *No.2* was the only locomotive in service for some time while traffic was very low. By March 1946, Baldwin locomotive *No.4* was permanently out of service, and George Edwards fitted its steam-driven water-lifter to *No.2* to save himself the effort of hand pumping.[30] Nevertheless, in the autumn a set of new steam pipes for *No.4* had been ordered from W. G. Bagnall and received on 6 November 1945. *No.3* was fit for service, and in late June 1946 permission was asked of Austen to put this locomotive back in steam, as *No.2* had failed on 21 June. In a memo of 25 June, Austen stated that he had not intended the Baldwins to be left unused indefinitely, as their condition would have seriously deteriorated. Then he instructed that, when made serviceable, the locomotives were to be used in weekly turns, and asked why Edwards had not fitted the new steam pipes to *No.4*. With the exception of the steam pipes, Bill Austen knew of no reason that *No.4* should be out of service. Reading between the lines, Edwards might have seen good reason for himself, which was shortly to be confirmed.

Will Jones retired on 3 July 1946, and Jackie Belter took over what remained of his duties. That same month the locomotives were condemned by the boiler inspector.[31] As early as 27 June, a horse and driver from E. L. Rogers of Malehurst Farm, Minsterley were hired for the haulage of empty wagons up to Callow, replaced

No.2 under the Halvans bridge on 17 July 1940, left there by the fitter-driver for the following shift.
J. G. VINCENT

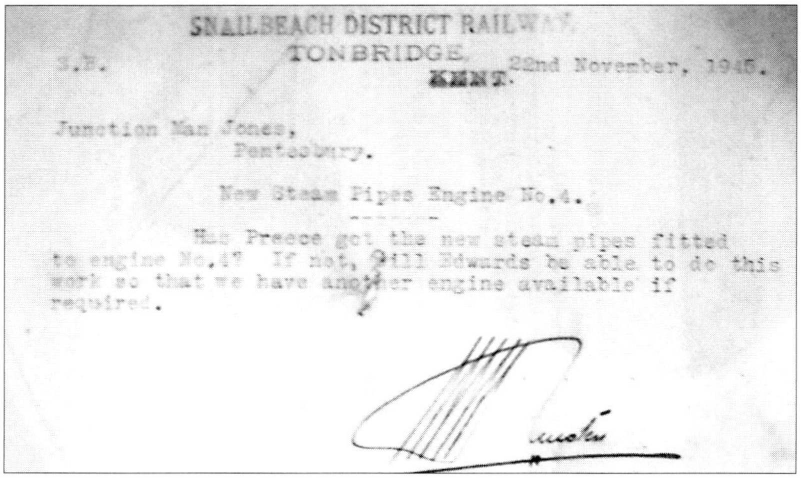

SNAILBEACH DISTRICT RAILWAY,
TONBRIDGE,
KENT.

S.B. 22nd November, 1945.

Junction Man Jones,
Pontesbury.

New Steam Pipes Engine No.4.

Has Preece got the new steam pipes fitted to engine No.4? If not, will Edwards be able to do this work so that we have another engine available if required.

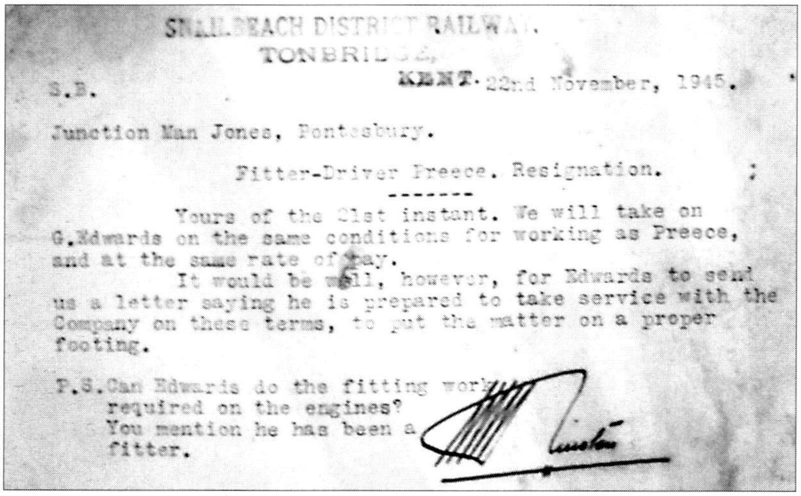

SNAILBEACH DISTRICT RAILWAY,
TONBRIDGE,
KENT.

S.B. 22nd November, 1945.

Junction Man Jones, Pontesbury.

Fitter-Driver Preece. Resignation.

Yours of the 21st instant. We will take on G.Edwards on the same conditions for working as Preece, and at the same rate of pay.

It would be well, however, for Edwards to send us a letter saying he is prepared to take service with the Company on these terms, to put the matter on a proper footing.

P.S. Can Edwards do the fitting work required on the engines? You mention he has been a fitter.

Two memos from Bill Austen in 1945 about locomotive maintenance and the possibility of employing George Edwards as a fitter-driver.
COLONEL STEPHENS RAILWAY MUSEUM ARCHIVE

on 8 July by a farm tractor, registration number BUX 174, owned by A. E. Davies. In July, the most empty wagons the tractor was able to haul was recorded as 20 spread over four trips. In September 1946, Austen instructed George Edwards to drain all water from the locomotives as a frost damage precaution. He might have thought optimistically that it was possible to find a buyer for them, other than a scrap metal merchant. It turned out not to be the case.

H. F. Stephens and his officers succeeded in resurrecting the railway, in an era when road development was flourishing and even killing off what trade the railway had left at Snailbeach. This was because the SDR controlled the only access to a quarry supplying good quality road stone. Paradoxically, a steam-hauled railway was making modest profits when road transport was becoming increasingly popular, by shipping the road metal which made it possible. With the Second World War, there was simultaneously a fall rather than a rise in demand for the railway's only freight, substantial wage inflation, and a general wearing out of the locomotives. The railway survived the war, again with no reserves for re-investment. But it still had a monopoly of access to a commodity which was to become more valuable than ever. The final chapter of the railway's operating history was about to be written by its client, the Shropshire County Council.

CHAPTER NINE

Road stone and rubber tyres 1947–1985

Early in 1947 the railway company intimated that it would cease operating for financial reasons, but until an alternative was agreed would continue the service using the hired tractor. At the same time there was a temporary agreement whereby the Shropshire County Council could operate its own traffic using the tractor on the company's track. The railway was still vital to the operation of the quarry. A full operating agreement was then made, which began on 14 April 1947 and was renewable annually.[1] A written agreement followed dated 8 April 1948, stating that the SDRCo could not economically maintain the line and giving the SCC powers to run it themselves.[2] The SCC was to have use of the track between the quarry and Pontesbury sidings and the wagons, and keep everything in a reasonable state of repair. It would also supply the haulage power and labour. It already paid a yearly rent of £12 for land, weighbridge and sidings, which would be merged into the new charge of 4d per ton for all traffic, with a fixed minimum payment of £100 per annum. The SCC was to permit the railway company to work any other traffic which might arise on this section of the line, but subject to the SCC's own traffic not being prejudiced. It was also to indemnify the company in respect of all third party liability regarding this section of the line. The indemnity was needed because the SDR was legally a common carrier and not able to exclude other traffic. The purchase of a dumper tractor for the railway haulage was authorised. But haulage continued to be done using the Fordson tractor driven by George Edwards, a County Council-employed fitter (see Chapter Eight). Of the railway staff, Will Jones had retired in 1946, leaving only Jackie Belter, who transferred to the SCC and continued to work with George on the railway (see Chapter 15).

There was now no further use for a junction with the Minsterley branch, because no steam coal was required. The railway company asked that the Pontesbury siding be put out of use to save maintenance charges, and accordingly it was 'spiked over'. During the post-war nationalisation of railways, the British Transport Commission (BTC) put every minor railway in Britain under the microscope. In June 1948 the regional head of the Railway Executive told the BTC that there was little prospect of the stone traffic on the SDR being transhipped to the branch line, so did not

A train of empty wagons for Callow Hill quarry hauled by the Fordson tractor at Pontesbury, c.1950. Driving is George Philpott Edwards with John Arthur Rowson (Jackie Belter) standing. The bridge plate reads 'Brymbo 1875'.
COURTESY *SHREWSBURY CHRONICLE*, A. YAPP & E. HUMPHREYS

Snailbeach mine 1 May 1949, recording a visit by enthusiasts. B. ROBERTS, COURTESY INDUSTRIAL RAILWAY SOCIETY

recommend the railway for acquisition.[3] Nationalisation spelt the end for most of what remained of H. F. Stephens' old railway empire, and the Salford Terrace offices closed down on 5 June 1948. Now that the SDRCo was little more than a landlord, it could run its affairs easily from smaller premises elsewhere.

With the operating agreement recently made with the SCC, the company decided to invite tenders for scrapping the unused part of the line. The locomotives were scrapped by Thomas W. Ward of Sheffield in May 1950.[4] The track was lifted above Callow Hill quarry to Snailbeach in the summer of 1951.[5] The company sent a lorry to take away the weighbridge equipment, but the operating levers remain encased under the road at Snailbeach to this day.[6]

Looking towards Minsterley at Pontesbury, c.1950. George Edwards driving with Jackie Belter standing. There is also a bridge plate this side. Photographs of Jackie in Chapter Fifteen are of the much younger man. COURTESY *SHREWSBURY CHRONICLE*, A. YAPP & E. HUMPHREYS

In 1953 the District Nurse Charlotte Hand died.[7] The shed on the Snailbeach railway wharf which she had used as a garage then fell vacant, and Crowsnest resident Douglas Challinor took a lease of the wharf for his car-breaking business. The board became acquainted with him because of this, and he was asked to act as driver during the annual inspection of the line. On these occasions, the directors and secretary stayed at the Britannia Hotel, Shrewsbury, having travelled by train to Shrewsbury on the Friday evening. They were collected by Doug on the Saturday morning and driven to Snailbeach. Having made an inspection they moved on to Callow Hill quarry and then either walked down the line or were driven to Pontesbury for an inspection there. They then returned to Shrewsbury, where the entire party had lunch before the board members returned to Tonbridge. Doug was also roped in as the company's resident agent, and was asked to report any problems concerning the railway property, which was principally the track bed and other land.[8]

On the death of Bill Austen in 1956, Arthur Iggulden acquired his shares and became chairman. Some shares were then passed to his sons Michael and Anthony Iggulden, and Michael was then appointed a director. Anthony

One of the Baldwin locomotives being cut up in 1950 outside the locomotive shed. The large gas bottle can be seen between the tracks. To the right are pieces of the plate frames of the Kerr Stuart locomotive *No.2*. J. E. SIMPSON COURTESY P. INGHAM

was also a director for a short time. Also in 1956, the SDRCo agreed that it would lease to the Council the track bed of the railway between Callow Hill quarry and the Callow Lane road bridge for the haulage of stone.[9] On 12 October 1956, Hayward's Quarries sold to the SCC the freehold of Callow Hill, Poles Hill, Nill's Hill and Resting Hill. With the sale of their property, Haywards Quarries Ltd decided to liquidate, and completed voluntary winding up at an extraordinary general meeting on 24 July 1957.[10] The promised lease was made by the SDRCo to the SCC dated 25 July 1957, for 21 years from 1 January 1957, allowing a through road for lorries from the Callow Lane to Pontesbury to be built. The minimum yearly rent was £150. A rate of 6d per ton was payable for carriage of more than 6,000 tons per annum, up to and including 12,000 tons. Over 12,000 tons, up to and including 20,000 tons per annum the rate was 4d per ton; and over 20,000 tons, 3d per ton. Traffic on the railway continued until 1959. The accommodation bridge over the railway at Woodhouse Farm was rebuilt with steel joists and

The tractor parked to one side at Pontesbury, 16 December 1951. B. ROBERTS, COURTESY INDUSTRIAL RAILWAY SOCIETY

Looking north at Callow Hill about 1950, showing the point to the private siding on the right with the tractor and a wagon. A. BARRELL

concrete. When making this lease, it was also agreed to deal with the three old bridges over the disused upper line, which were declared in bad condition. The company transferred the bridges to the SCC under the Bridges Act 1929. The SCC was to fill in the arches, and in the event of any decision to re-open the line, the company was to reimburse the County Council for any work needed as a result. In 1959, the two bridges at Snailbeach were duly filled in.[11] The Callow Lane bridge was filled in, the parapets were removed and the lane was straightened. The iron plate bridge under the line at Pontesbury remained, until the housing development next to the railway was being built. The bridge had to be removed after it was hit by a tipping lorry with its body in the up position.[12] It was demolished in January 1969.[13] The stone abutments were removed, and the railway embankments cut back from the road and grassed over.

By 1960 increased output was reported as a result of road access to the quarry. In early 1960 it was decided that the oil engine-driven plant at Callow Hill needed replacing, and that the quarry should be re-equipped with electrically-driven plant. The total cost was estimated at £53,000. Goodwin & Barsby of Leicester equipped the quarry for £45,005 and the SCC Roads and Bridges Dept performed electrical and civil engineering work. In the winter of 1961 to spring 1962, the track between the quarry and Pontesbury sidings was lifted by the Wellington Coal & Iron Co. Ltd, allowing the final section of track bed to be metalled for lorry access. Some of the rail was bought by the Talyllyn Railway, a preserved narrow gauge railway in mid-Wales.[14] Installation of the new plant at Callow Hill was also completed. Developments continued in 1963 and 1964, including a new water main.[15]

3 September 1955. Looking north at Callow Hill, the main line is to the left and the private siding is to the right. A. J. PRATT

3 September 1955. End of the line just south of Callow Hill quarry looking north – the limit of track lifting in 1950. A. J. PRATT

The lease of the lower part of the line to the Shropshire County Council was finally renewed for a term of 21 years from December 1977, at an annual rent for the first seven years of £200. Wayleaves crossing the route provided some income. There were several with the Midland Electricity Board for overhead power lines. Tolls charged for the carriage of stone and other goods over the company's land totalled between £1,300 and £1,500 in the early 1980s.[16] Arthur Iggulden died in 1979, leaving his shares in the company to his family. Michael Iggulden then became chairman and his wife Susan a director. Shortly afterwards Michael bought his brother Anthony's shares.

In early 1984, the Snailbeach District Railways Company was offered for sale, with notice given in the press.[17] The Sale Prospectus invited offers in the region of £25,000 for the company as a going concern. The chief asset of any value was the land – mainly the track bed – but there were also wayleaves. Doug Challinor bought the freehold of the land and existing wayleaves above the Callow Lane and up to Snailbeach. Doug only wanted the wharf at Snailbeach, which he was still renting. On taking possession, he gave free to the Methodist church the piece beside the chapel and burial ground. To the village hall trustees he also gave free the piece beside the village hall down to the Halvans

Wagons awaiting scrapping in 1962. J. MILNER

bridge. Other neighbours of the railway bought what was useful to them. As well as the wharf, the piece from the wharf up to the mine was retained by Doug Challinor, but it remains in public use as a footpath. The freehold of the lower portion of the track bed was sold to the SCC in February 1985 and continued in use as quarry access.[18] This liquidation of the major asset marked the close of business for the railway company. But having been created by an Act of Parliament, the Snailbeach District Railways Company still exists, although it has no assets and does not trade. It would take another Act of Parliament to dissolve it, so it has been left dormant.

Effectively, the railway had died in 1946 when the locomotives were declared unsafe. Rubber tyred transport began to ease its way in then, with the use of the Fordson tractor, but the

Mid-1950s. Wagons at Pontesbury with *'1/8 th-to-dust'* road stone screenings dumped on the track. This was probably done to compensate for settling of the fill on which the siding over the lorry-loading bay was built.
J. M. LLOYD
COURTESY I. M. LLOYD

Pontesbury in early 1962, during the track lifting which made way for the County Council's lorries.
I. J. BROWN

At Callow Hill quarry in 1963, the Royal Daylight diesel oil tank has been removed from the old Bagnall wagon onto a plinth after track lifting. Mrs Iris Brown in attendance.
I. J. BROWN

Looking south with recently installed road stone plant at Callow Hill quarry in 1971.

Mine, railway and people: Snailbeach residents pose casually in Sunday best on a flat wagon outside the loco shed, with Snailbeach Coppice in the background. COURTESY A. YAPP.

railway company kept the upper hand. It had what the quarry operators needed. The poor road access to the Stiperstones gave rise to the railway in the first place, and even after the mines failed, the line was the most practical way of removing the new wealth of the hill: stone and timber in remote places. As before, the company managed by removing old disused sidings here, and building new, useful sidings there. In the final flourish of adaptation, the company removed all of the track, and traded for another quarter century by embracing the age of the heavy lorry. This needed no capital investment by the company but attracted modest revenue from the wayleaves. In a way, the replacement of the rails with a road acknowledges the effectiveness of Henry Dennis's scheme. The railway had outlived its usefulness, but even for the new transport of choice, its route could not be bettered. Even Dennis himself might have approved.

CHAPTER TEN

Engineering, way and works

COMPARATIVE COSTS PER MILE OF LIGHT AND NARROW GAUGE RAILWAYS

The estimates for the various competing schemes in the lead field certainly indicate that the SDR scheme was economical (see Chapter Four). Data in the table below compares the SDR with lines actually built elsewhere. Figures for the construction costs of other railways of the period illustrate the moderate cost of the SDR. This is despite the abandonment of Railway No.2, of only 1 mile 7 furlongs 3.90 chains in length, and the spending of all the capital on Railway No.1 alone, having a length of 3 miles 2 furlongs and 5.44 chains. It had moderate amounts of earth and rock works. The Talyllyn Railway had little in the way of civil engineering works, accounting for its low cost. The Culm Valley Light Railway had no major engineering works, but was standard gauge and had passenger and goods facilities. An added cost factor in the Snailbeach line was that the earthworks were wide enough to take a standard gauge line later if required, although it must be remembered that labour was relatively cheap.

If we also compare the cost of the SDR with the estimate for Kough's first South Shropshire Railways scheme, we have a valid local comparison, because it had a similar route direct from Pontesbury to Snailbeach. The estimate for Kough's line was £6,000 per mile. Looking at the deposited plan and section, this seems to be a gross underestimate. Dennis's actual cost of construction was about the same as the estimate for Kough's scheme, yet the latter had vastly greater earthworks. It was planned to take a course a little further east of Dennis's, in terrain which was difficult to navigate and expensive to excavate in the body of the Stiperstones. The scheme would have greatly overspent. By all measures, Dennis's Snailbeach line was a well engineered and economical solution.

THE ROUTE, STATION LAYOUT AND FACILITIES

Pontesbury Junction or Bottom Wharf

In its original form, there were four narrow gauge sidings at Pontesbury which fanned out and interlocked with the standard gauge facility.[2] There was a fifth siding joining one of these, for loading onto road vehicles. Within the company's boundary, the sidings of both gauges were owned and installed by the SDR. All the other standard gauge installation – the junction itself – was carried out by the Joint Railways Committee's engineering department at the SDR's expense. The junction and sidings were planned by Henry Dennis, with the approval of the Joint Railways committee as far as its operating interest was concerned (see Chapter Five). Junction equipment included interlocking post and disc signals for the single track main line and siding points, with everything operated from a cabin, and a two-way gong communication between the junction & Pontesbury station. The cabin, signals, patent locking apparatus, connection rodding and all necessary hardware were supplied by McKenzie & Holland for £362 13s 4d.[3]

On 15 October 1926, *The Locomotive* magazine reported that '*The layout of the transfer*

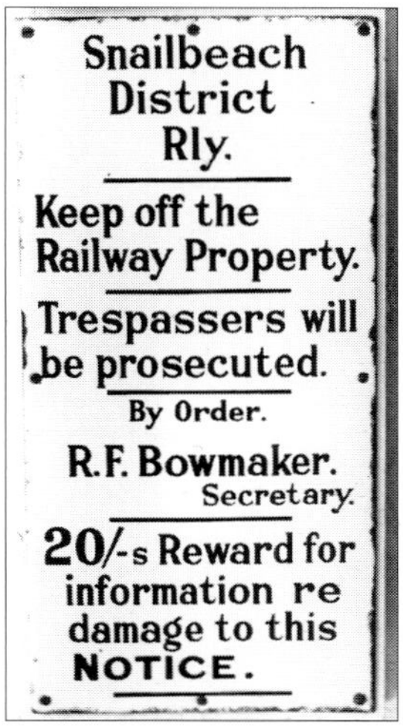

The standard post-1922 No Trespassing sign.
V. J. BRADLEY

TABLE OF COSTS[1]

Railway	Year opened	Gauge (feet-inches)	Route miles completed (miles-furlongs-chains)	Approx. cost per mile
Culm Valley Light Railway	1876	4 – 8½	7 – 2 – 0	£6,300
Talyllyn Railway	1865	2 – 3	6 – 5 – 0	£2,150
Snailbeach District Railway	1877	2 – 4	3 – 2 – 5.44 from Engineer's Estimate of Expense	£6,150

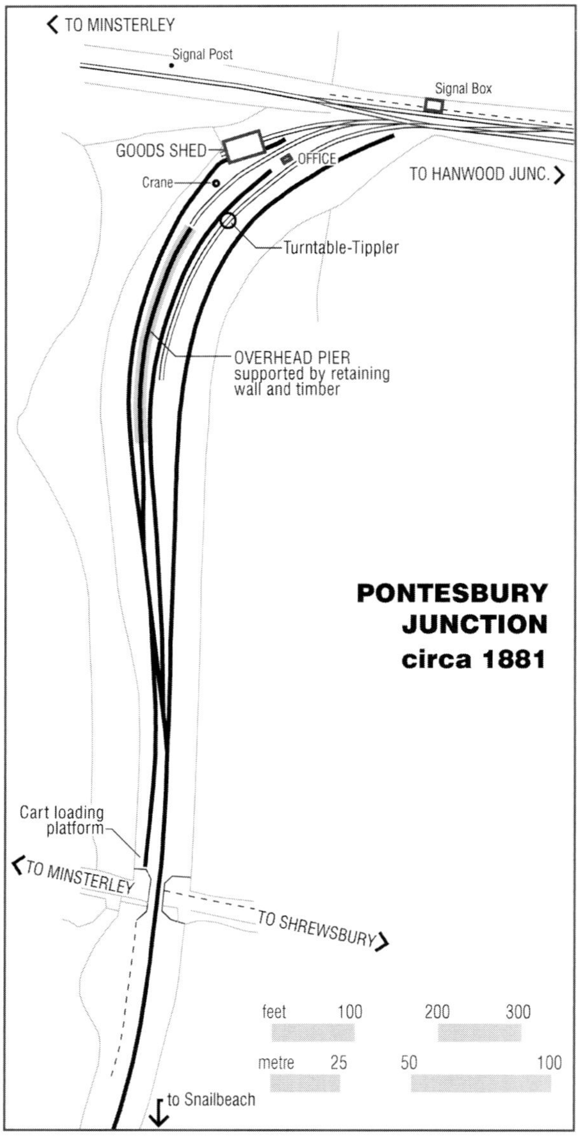
TO MINSTERLEY
Signal Post
Signal Box
GOODS SHED
OFFICE
TO HANWOOD JUNC.
Crane
Turntable-Tippler
OVERHEAD PIER supported by retaining wall and timber
PONTESBURY JUNCTION circa 1881
Cart loading platform
TO MINSTERLEY
TO SHREWSBURY
feet 100 200 300
metre 25 50 100
to Snailbeach

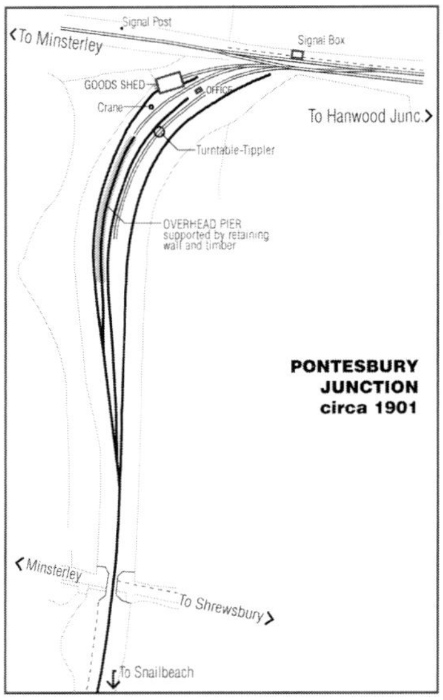
Signal Post
To Minsterley
Signal Box
GOODS SHED
OFFICE
Crane
To Hanwood Junc.
Turntable-Tippler
OVERHEAD PIER supported by retaining wall and timber
PONTESBURY JUNCTION circa 1901
Minsterley
To Shrewsbury
To Snailbeach

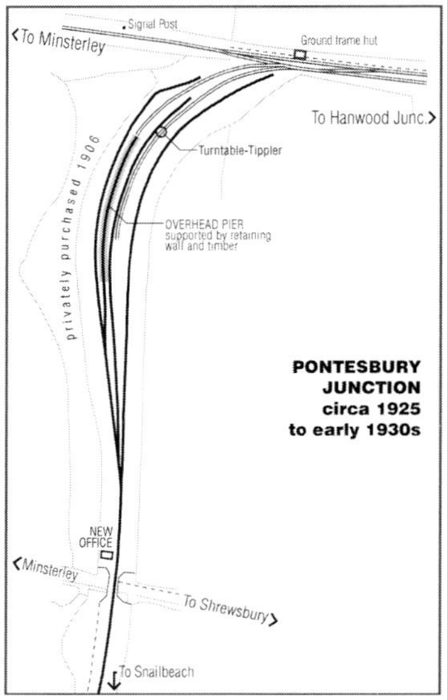
Signal Post
To Minsterley
Ground frame hut
To Hanwood Junc.
Turntable-Tippler
privately purchased 1906
OVERHEAD PIER supported by retaining wall and timber
PONTESBURY JUNCTION circa 1925 to early 1930s
NEW OFFICE
Minsterley
To Shrewsbury
To Snailbeach

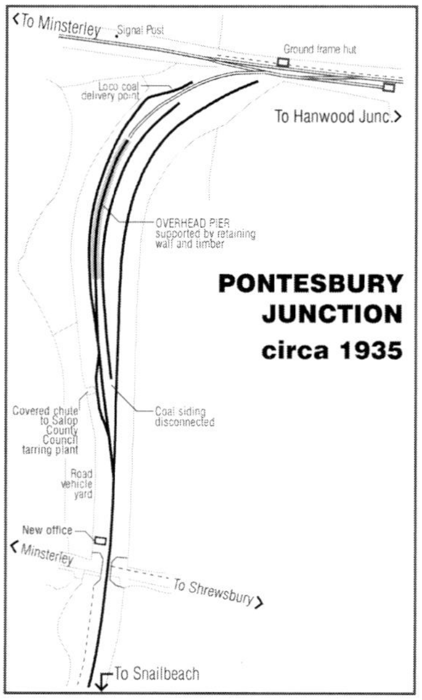
To Minsterley
Signal Post
Ground frame hut
Loco coal delivery point
To Hanwood Junc.
OVERHEAD PIER supported by retaining wall and timber
PONTESBURY JUNCTION circa 1935
Covered chute to Salop County Council tarring plant
Coal siding disconnected
Road vehicle yard
New office
Minsterley
To Shrewsbury
To Snailbeach

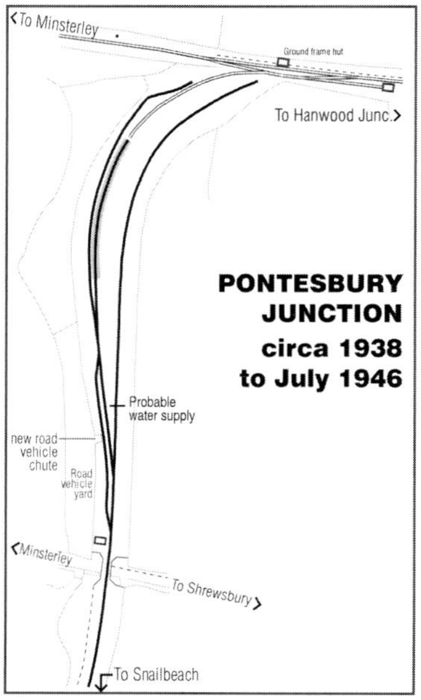
To Minsterley
Ground frame hut
To Hanwood Junc.
PONTESBURY JUNCTION circa 1938 to July 1946
Probable water supply
new road vehicle chute
Road vehicle yard
Minsterley
To Shrewsbury
To Snailbeach

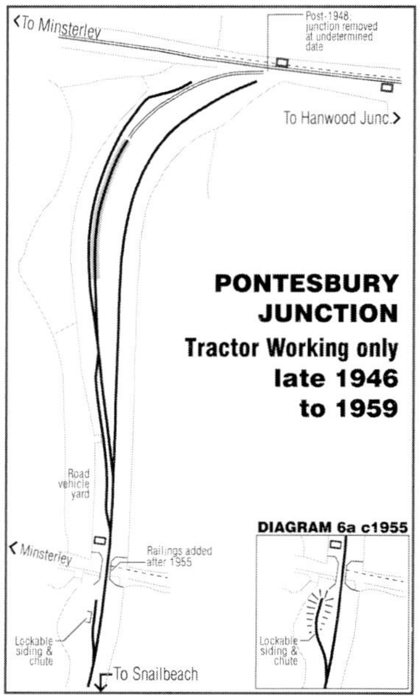
To Minsterley
Post-1948 junction removed at undetermined date
To Hanwood Junc.
PONTESBURY JUNCTION Tractor Working only late 1946 to 1959
Road vehicle yard
DIAGRAM 6a c1955
Minsterley
Railings added after 1955
Lockable siding & chute
To Snailbeach
Lockable siding & chute

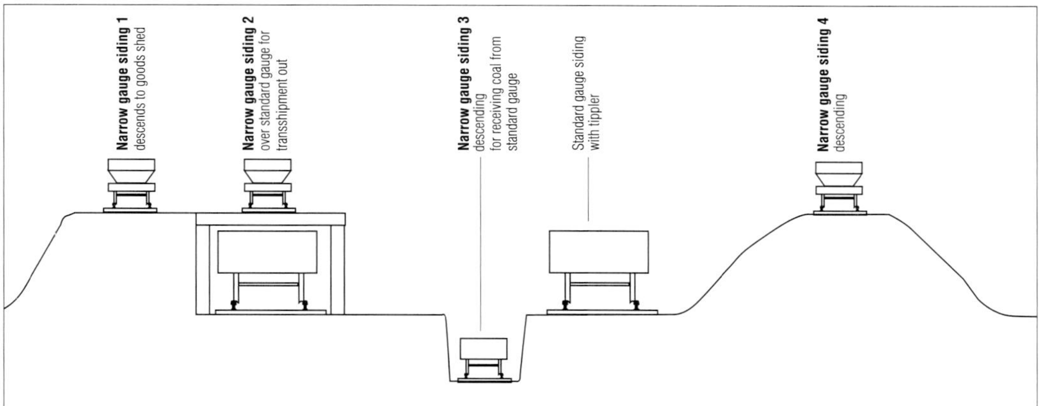

Schematic cross-section of sidings at Pontesbury showing transshipment arrangements (not to scale)

yard at Pontesbury Junction, from narrow gauge to standard gauge, and standard to narrow, ... although built over fifty years ago, is decidedly good, and reflects great credit on the designer.' It then described how gravity and machinery was used, and this is also illustrated by Dennis's working plans. Transhipment of minerals was made as easy and economical as possible by constructing the sidings at different levels and on gradients. This allowed gravity unloading between wagons positioned one above the other (see cross-section) and moved along by hand. In part, this arrangement was made possible by the line entering Pontesbury on an embankment, allowing the road to be bridged. The embankment was continued into the sidings, giving height advantage. Heavy goods, for example timber and pig lead, were transferred using hand-operated cranes, supplied by W. H. Hughes of Plaskynaston Foundry, Ruabon.[4]

Narrow gauge siding 1, on the west side, descended towards the standard gauge branch line and passed through a timber-built goods shed. This was shared with a standard gauge siding on the opposite side of a platform nine feet wide having a one-ton crane. This shed was principally for the transhipment of pig lead, but any goods to be kept dry and secure will have passed through it. There were doors at both ends on both tracks. Between 1901 and 1907, the goods shed was removed. No pig lead was being shipped so the facility was redundant.

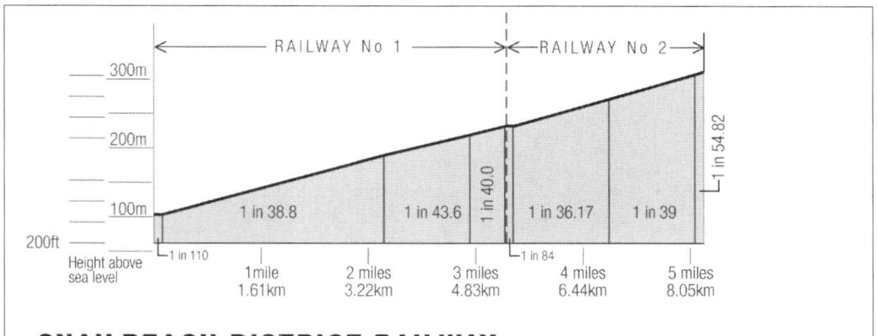

SNAILBEACH DISTRICT RAILWAY

GRADIENT PROFILE FROM DEPOSITED PLAN

Siding 2 turned off No.1 and ran onto an overhead pier with a standard gauge siding below. This was for the outgoing shipment of metal ores, barytes and spar, discharged from the bottom of hopper wagons. The horizontal baulks of the pier had a cross-section of about ten inches by twelve inches, supported on the west side by a massive stone wall.[5] This was also a retaining wall for the heavy earthwork on which siding 1 was laid. The pier was lengthened for the Granham's

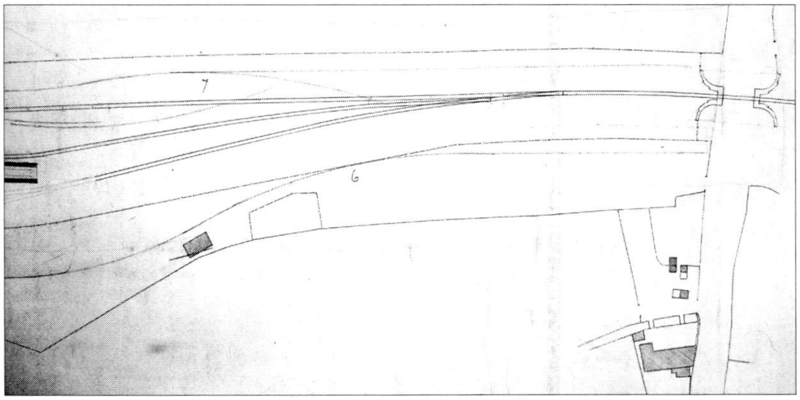

Extract of working plan for sidings at Pontesbury. The entry point for siding 3 (the coal siding) is shown on siding 1, as seen in all known photographs, rather than on siding 4 as shown by the Ordnance Survey County Series 1st edition 1:2,500 scale map. The entry point for siding 2 (the narrow gauge to standard gauge mineral exchange siding) is not shown, but again is known to have been on siding 1. The ends of the parallel stone walls for this appear at the far left. The County Series maps showed this point on the erroneous siding 3. This would have been impractical for operation, because of the dip between the embankments carrying sidings 1 and 4.
SHROPSHIRE ARCHIVES 4743/48

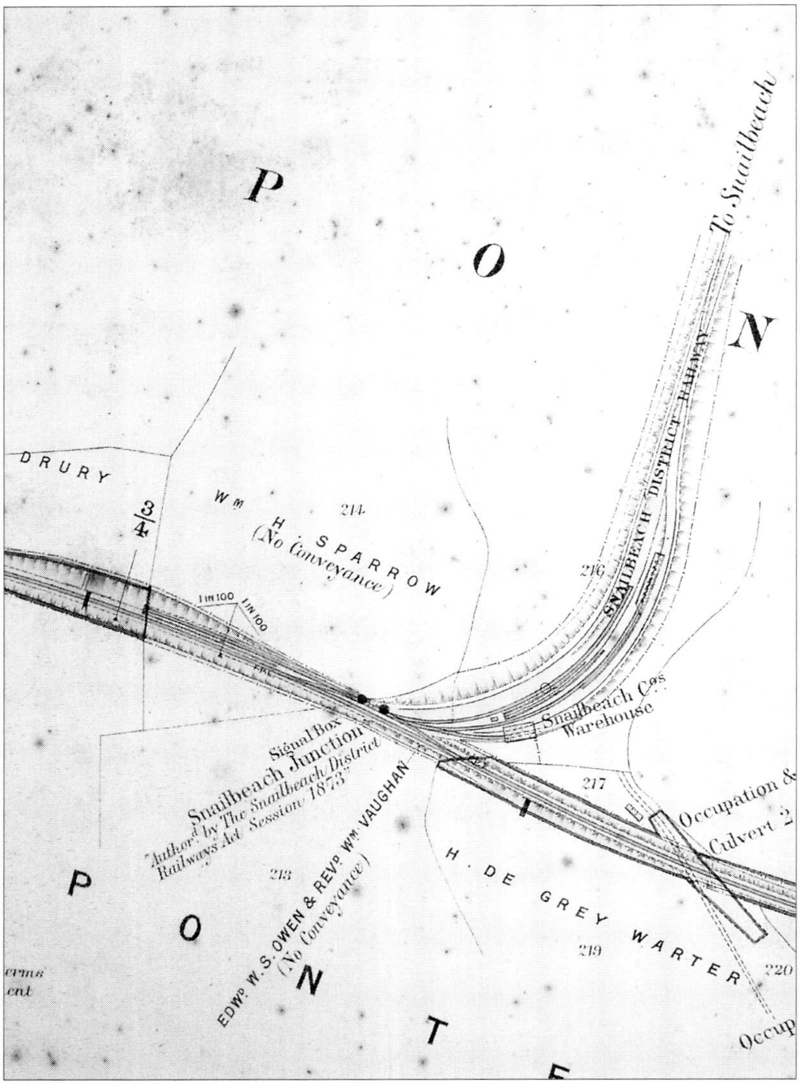

Shrewsbury & Welshpool Railway plan of Pontesbury Junction, 1886. The surveyor made a mistake with the entry point for siding 2. M. J. SHAW

Moor quarry traffic.[6] The same standard gauge siding was for transhipment of heavy items to and from siding 1, for which a four-ton crane was provided which swung between the two lines.[7] The principal traffic here was timber for the mine. This crane was broken by the early 1890s and it is not known if it was repaired. By the mid-1920s a derrick had been installed in about the same location as the goods shed. This was used for loading timber, which was by this time an outbound traffic (see Chapters Seven and Eight).

A plan found in the National Library of Wales dated spring 1875 shows that a narrow gauge turntable-tippler was proposed originally for transshipping minerals. The overhead pier plan improved access to the goods shed and made room for the 4-ton crane

Siding 3 descended below the level of the standard gauge sidings, and terminated in a rubble-walled cutting. This siding was for loading coal wagons, to go to the smelt works and the mine. The next standard gauge siding had a turntable-tippler installed. This was for unloading coal wagons with end doors. Each wagon in turn was passed onto the machine and turned through 90 degrees. Using a geared handle, it was then elevated a few degrees to discharge coal into a narrow gauge wagon waiting in the cutting. Several narrow gauge wagons were needed to take one standard gauge wagon-load of coal. There was a wooden platform set in the cutting opposite the tippler for supervising the operation. Coal traffic was so small after 1925 that the turntable-tippler was soon removed, and any good rails from the coal

Pontesbury, 8 October 1939, showing the recently constructed lorry-loading bay and the earlier tarring plant. The accommodation of the new siding to the loading bay is notable. The rails are spiked directly to long horizontal timbers which rest on the original embankment and the curtain walling structure seen here. The space between is filled level with quarry waste. J. G. VINCENT siding used elsewhere. Locomotive coal and the little coal which was re-sold was then delivered by shovelling into wagons standing on siding 1.

Close to the coal siding cutting was a tiny timber-built office. By the 1920s this had been replaced with another office close to the Minsterley-Shrewsbury road. This was convenient for meeting the increasing motor traffic and connecting to the national telephone system. In later years, a two-tank water supply system for the locomotives was built at Pontesbury (details are unknown), with an overhead feed pipe which appeared around 1940.[8]

Siding 4, on the east side, descended steadily towards the standard gauge railway. It was a spare siding and used for wagon storage. Allowance was made for converting the railway to standard gauge, and in that case this siding would probably have become the main line. According to Dennis's plan of May 1876 submitted to the Joint Railway engineers, the siding was marked 'Narrow Gauge Passenger Line', and it was intended to build a passenger platform area beside it, close to the branch line. Also planned was a paved footpath to Pontesbury station. Neither was built. The footpath was to go along the side of the standard gauge railway cutting, and the Joint Railways committee probably had grounds to object. The plan also indicates another footpath running south across the fields from the SDR platform, presumably to meet the main road. No footpath is shown to have existed there by the 1881 Ordnance Survey. This was to cross private land, so creation of the path would have incurred a wayleave charge. Dennis was probably recording later intentions. The company's Prospectus of 1875 indicated the intention to introduce a fare-paying passenger service later, which the Act of Parliament authorised, but it never happened.

Siding 5 trailed south off siding 1. It descended several feet to serve a wooden platform for loading coal and lime onto road vehicles. The loading point was beside the main road from Minsterley to Shrewsbury, and close to the bridge. This siding was lifted in the late 1890s. A new siding off siding 1 was built for the Shropshire County Council (SCC) after 1933 for feeding road stone to a tarring plant from SDR hopper wagons. Full wagons from Callow Hill quarry were run down to it by gravity. The field to the west of the sidings, bought in 1906, was fenced into two parts, and the south portion rented to the SCC for a road vehicle yard and a site for the tarring plant. South of the road, the railway land on the west of the embankment was also included in the rental, and a weighbridge was installed. Both north and south of the road, road stone was shovelled out of box wagons and down the embankment, at these points known as the 'chutes'. By September 1938, a new loading platform was built for loading untreated road stone from hopper wagons into SCC lorries, and the tarring plant siding was extended to serve it.[9] This replaced the 'chute' north of the road.

Pontesbury to Callow Hill

As the line climbed out of Pontesbury sidings, it crossed the Shrewsbury to Bishop's Castle road (A488) on an iron plate bridge. Mounted on masonry piers, it was made by the Brymbo Iron Works and was wide enough to take a standard gauge track.[10] Directly after the bridge on the

In the cutting between Pontesbury and Woodhouse Farm, looking south. 26 July 1952.
F. W. SHUTTLEWORTH

Woodhouse Farm accommodation bridge looking towards Snailbeach. 26 July 1952.
F. W. SHUTTLEWORTH

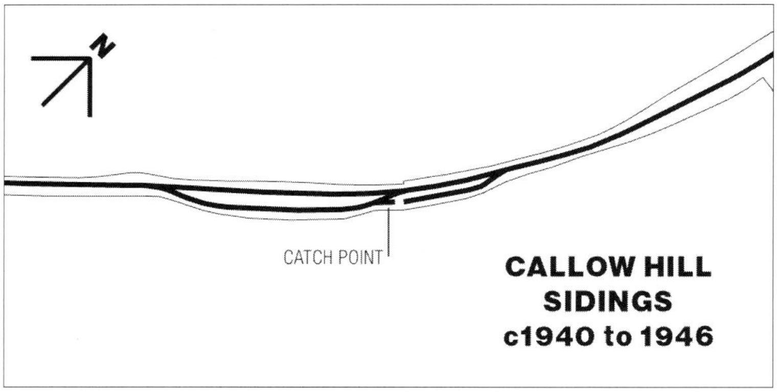

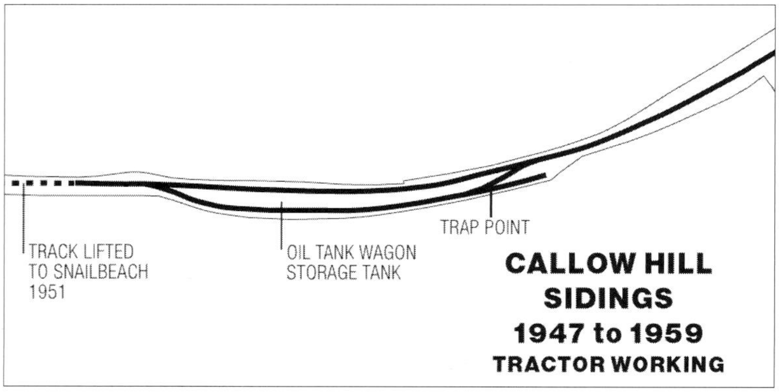

west side, the 'chutes' of the 1930s were passed, and a short siding was installed in the era of tractor haulage (post-1946) for dumping unwanted screenings. The line entered a cutting, and before long passed under the approach road to Woodhouse Farm. This crossed the railway on a single timber-built span mounted on stone piers.[11] Soon after, the line emerged from the cutting. From here to the Callow Lane there were few notable features until the opening of Callow Hill quarry by S. J. Hayward in about 1924. Before 1900, the Callow Hill lead mine was worked sporadically. Ore was taken up to Snailbeach from here for dressing, but no wagon loading facilities are known of.[12] When the Callow Hill quarry traffic began, a passing loop was soon laid by the SDRCo to serve it. Between 1927 and 1930 Hayward's Quarries Ltd installed a longer private siding with a weighbridge to serve new machinery directly. This was joined to the main line by crossovers installed by the SDRCo, and the old passing loop was removed. There was also a locomotive watering point.[13] On Hayward's ground close to the quarry's later Ruston Bucyrus engine was a short siding for a fuel oil tank wagon. In the early 1940s, another siding was laid north of the loop.[14] The sidings changed again after steam haulage ceased.[15] The needs were different when using a farm tractor to haul and marshal the wagons.

Callow Hill to Snailbeach Farm

Skirting Callow Hill, the line was built on a shelf overhung by the wooded hillside. It then crossed the most significant stream en route by way of a deep embankment and a culvert. As the line emerged above the Callow Lane, the platelayer's smithy was positioned on the east side and opposite the road. In one corner it had the traditional pattern smith's hearth with a bellows, an example of which has been rebuilt in the blacksmith's shop at Snailbeach mine. The lane quickly climbed an embankment and over the

The derelict railway smithy, 3 September 1955, with the traditional hearth and bellows still in place.
A. J. PRATT

A postcard showing the skew bridge taking the Callow Lane over the railway, looking south towards Snailbeach. The sun is behind the man sitting on the parapet, hence the 'burnt-out' effect at the top of the picture.
COURTESY S. A. LELEUX

The line, Prospect House and the Big Chimney. This view is very early in the history of the railway. The ballast is clean and well graded, and the lineside fence is in very good condition. Dennis's specification details fence construction very clearly. There is also no hedgerow growth along the fence yet, although there is some low shrubbery. COURTESY MRS PEGGY CHIDLEY

railway on a skew bridge built from rough-hewn stone. The semi-circular arches of this and the two bridges at Snailbeach were lined and decorated with spiral-laid brick. By the 1950s a line-side wall was demolished at one point to allow access to the track bed.[16]

The railway was now accompanied on both sides by alternate woodland and fields, with minor earthworks and crossing several culverted streams on its way to Snailbeach Farm. After the track was lifted, the track bed was used as a cart track as far as the Farm Lane level crossing. A couple of hundred yards after the skew bridge, the line entered an S-bend. To the left of the second curve there was a marshy pond which in later years was used as a locomotive water supply. The water was hand pumped from this pond to locomotives *No.2* and *No.3*, whereas Baldwin *No.4* was fitted with a steam-driven water-lifter.[17] Part of a field to the left was leased from the Marquis of Bath by the Granham's Moor Quarry Co. on which their private tramway terminated. The SDRCo installed a turnout to connect with it, and the stub of siding remained after the tramway was lifted, allowing locomotives to reach the pond.[18] A little further on, the branch to the smelt works forked to the right and on a down gradient. A cast iron water column for the locomotives stood just before the turnout and on the right of the line.[19] There was also a tank to feed the water column. The mine managers' houses, Prospect House and West View, were passed on rising ground to the left. After the Second World War, the newly built Prospect Cottages were passed on the right, and Farm Lane was crossed on the level.

The felspar siding, Snailbeach, looking roughly south across the white hillocks. The track curving away to the right leads to the main line. The stepped wall arrangement created platforms for machines, to allow riddling of the gravel with the advantage of gravity and working towards the loading of the wagons. A. BARRELL

Snailbeach Farm to Snailbeach station

A few yards further on, the felspar siding which served the spoil heap workings (and was part of the spoil heap lease) was passed on the left.[20] The siding remained intact when the main line was lifted, but by the mid-1950s the rails were reported lifted to one side. The line then passed under the second of the bridges (known in living memory as 'the Halvans bridge'), taking a straight course from there to the terminus at the wharf. On the left, directly after the bridge, the

View south in the cutting at Snailbeach, with the Methodist chapel in the distance.

The Halvans bridge looking north in the 1930s. At the sleeper closest to the camera, notice the left hand rail joint where an old wrought iron rail (bottom) meets a new steel one – the profiles do not match, but trains can run over it safely at low speeds.
H. F. G. DALSTON

Snailbeach wharf looking north at Hughes's bridge, 15 September 1941.
R. K. COPE

Snailbeach barytes mill was built around 1890. Several similar businesses followed the first mill lessees.[21] In the 1920s a trailing siding was laid for the last of them, Gray's spar gravel works. The line then passed between terraces of cottages and Lower Works.

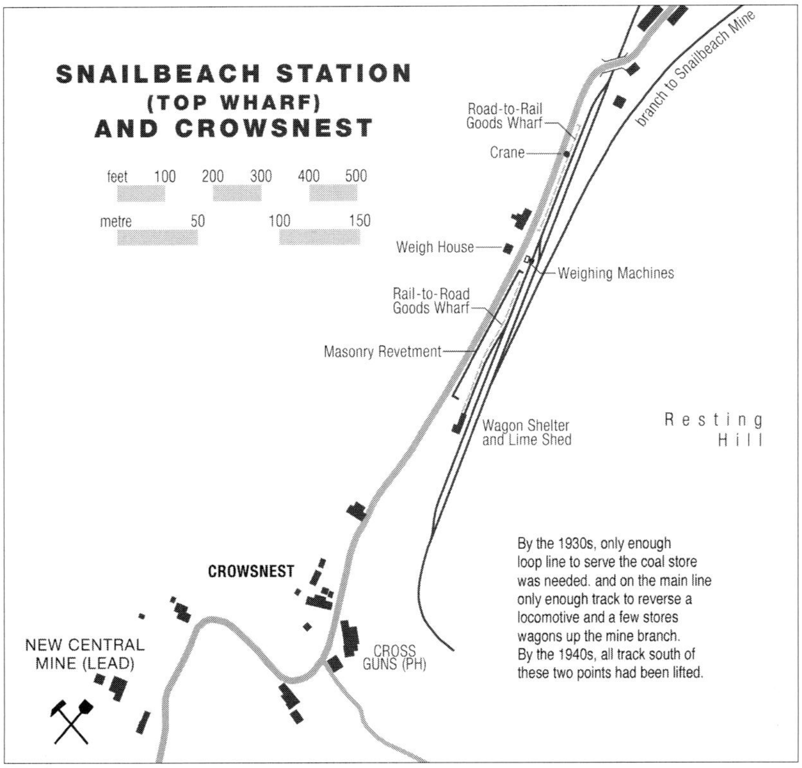

Snailbeach station (or Top Wharf)

After the Primitive Methodist Chapel and a pair of cottages, the railway passed under the final road bridge (known in living memory as 'Hughes's bridge' after the blacksmith Stanley Hughes, whose smithy still stands close by [22]) and entered Snailbeach station or Top Wharf. Like the other two bridges, it originally had stone parapets, but here they were replaced in the 1930s by iron railings. The bridge had become an accident blackspot, and removing the parapets improved visibility.[23] Immediately, a short passing loop diverged on the right hand side. The straight part of the loop was planned to be 145ft long, to take either 16 wagons or 14 wagons and a locomotive. Alongside the loop was a stone and concrete-built loading platform, again with a four-ton crane. The loop line continued after a crossover, and passed over a weighing machine. There was a cart weighing machine next to it, at the entrance to the wharf yard, and the balance arms passed in a conduit under the road to the weigh house. Here the weights were recorded and payments taken by the clerk for private sales later of coal, lime and building materials. The weighbridges and equipment were supplied by Hodgson & Stead of Salford. The cart weighing machine closed during the Great War.[24]

The main line, on the left, climbed sharply up

A postcard view of Crowsnest with Snailbeach wharf in the distance, before 1907. The larger house on the end of the terrace is the Cross Guns public house. COURTESY IRONBRIDGE GORGE MUSEUM TRUST

Looking up the bank to show the massive rough stone retaining wall, used to create the road vehicle yard at Snailbeach wharf by infilling

BELOW: The road-rail wharf at Snailbeach about 1925, with one of the Baldwin locomotives passing wagons at the crane, including a bogie wagon toward the camera, and a flat wagon in the foreground. The bridge still has its original parapet. COURTESY D. A. BREWER

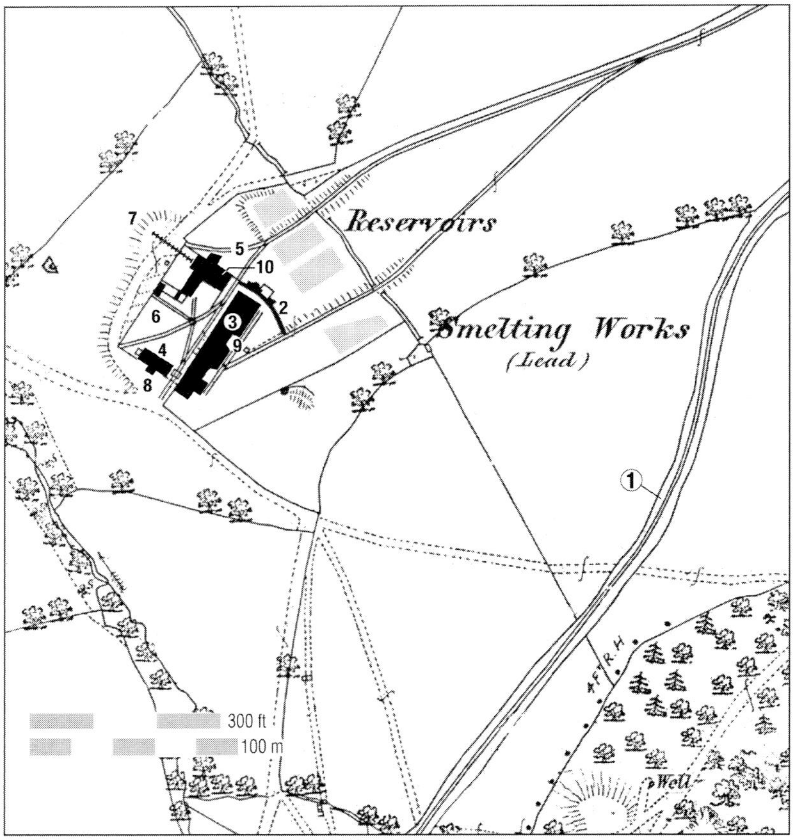

1 MAIN LINE OF SNAILBEACH DISTRICT RAILWAY
2 PART OF FLUE FROM SLAG HOUSE
3 FURNACE HALL
4 SMELT WORKERS BARRACKS (1st Fl.) and OFFICE & WEIGHBRIDGE HOUSE (Grd Fl.)
5 WASTE SIDING
6 SMELTING WORKS YARD (LEAD, FUEL & STORES SIDINGS)
7 WASTE HEAP & EARLY TRAMWAY FROM SLAG HOUSE
8 WEIGHBRIDGE
9 LEAD ORE SIDINGS
10 SLAG HOUSE

Directly relevant to smelting lead from ores, a Cost and Pay Book in a private collection gives the following wages payable on 7 July 1882 (the weights are in tons, hundredweights and quarters):
Samuel Purslow & Co Weighing Proprietors' Ores – Lead ores 157t-17-0 @ 2d £1 6s 4d
Six men removing ores and delivering to smelt works – 1 day @ 2s 9d, 16 days @ 2s 6d
Richard Davies & Co Loading Lead at Smelting Works

1st row	160t @ ½d	6s 8d
2nd row	20t @ 1¾d	2s 11d
3rd row	25t @ 2½d	5s 2½d
4th row	10t @ 3½d	2s 11d
5th row	5t @ 4½d	1s 10½d

to the junction for the mine branch. The loop line on the right continued essentially on the flat, with the road vehicle yard a few feet below, but then climbed again to rejoin the main line. On this loop line goods were unloaded from rail to road, including coal for other mines such as Tankerville. A short siding diverged and fell away slightly to a timber built shed with sliding doors, for storing lime. There was a canopy over the siding to keep full wagons of lime dry. By the early 1900s these two buildings were gone, but in the 1920s a rough roofless store was built in the yard a few yards north, to serve a small coal business. A roof was added when it was rented later as a garage.

The smelt works branch

This private siding divided part way down. The eastern fork delivered ore wagons to the back of the principal building of the works – the smelt house or furnace hall containing five reverberatory smelting furnaces. Ore was previously delivered here by the self-acting inclined tramway (see Chapter Two). The western fork entered the smelt works yard, where there were several sidings accessed by two turnouts including a triangle and a wagon turntable, to ensure the wagon doors were repositioned on the correct side of the line when turned round. Lead pigs were stacked in the yard before hand-loading onto wagons.[25] Much hand shunting would take place here. At the south corner of the works, there was a weighing machine placed in the road entrance. Ore from other mines in the area (and coal and other stores before the SDR was built) entered the works here.

The mine branch

The steepest stretch of track on the railway climbed quickly away from Snailbeach station and up onto the hillside, to pass behind the cottages. A summit was reached before the line dropped slightly and crossed Shop Lane.

A cutting was made for the line through the 'black hillock' of spoil, and bridge abutments were built to allow continued use of the spoil tramway. A fan of three sidings was served by a turnout right under the bridge. Two sidings served the locomotive shed, and the third served a concrete loading platform, butting right up to the wagons with a stone wall. The locomotive shed had a water supply, fed via pipe by Dennis's reservoir. The outlet was inside the building, and a large hose was attached, allowing the tank to be filled with the locomotive standing outside. But

View from the cab of *No.2* on the mine branch, 22 March 1942.
J. G. VINCENT

the water pressure was very low (certainly true after the Stephens takeover), and only the locomotives with their own water-lifting apparatus could take water here. The main branch line now ran around the north edge of a plateau of spoil which was the by-product of decades of mining. On this were the main ore dressing floors. A siding onto a dressing floor to the south of the main branch line was accessed using two points and a slip road of short radius to join a dressing floor tramway, which crossed the branch line using a right-angled crossing to take waste onto the tips and ore to a second crusher house. This arrangement was still *in situ* in c.1885, but was removed some time later. These were hand operations, or perhaps the waste needed to be horse drawn depending on tramway gradient.

Another siding turned south across the floors to serve the carpenter's yard. Timber was unloaded here using sheer-legs. The main siding continued, snaking between Black Tom engine shaft and Hewitt's cottage, and then to the head shunt near the magazine. The end of the head shunt was walled, and spillage there shows that it was used for loading wagons with minerals from carts. The Halvans Co. often carted dressed barytes from Perkins Level to the railway.[26]

From the head shunt, a reversing line climbed gently to the bottom of the hill, gaining a few feet in height. This entered a raised coal yard, built behind retaining walls, which served Dennis's new boiler house. Part of the yard was used by the SDR for locomotive coal. Before crossing the road to Lord's Hill, a further siding split off to serve coal wagons to a rope-worked incline. The incline was single tracked with a passing place, and was worked by a steam

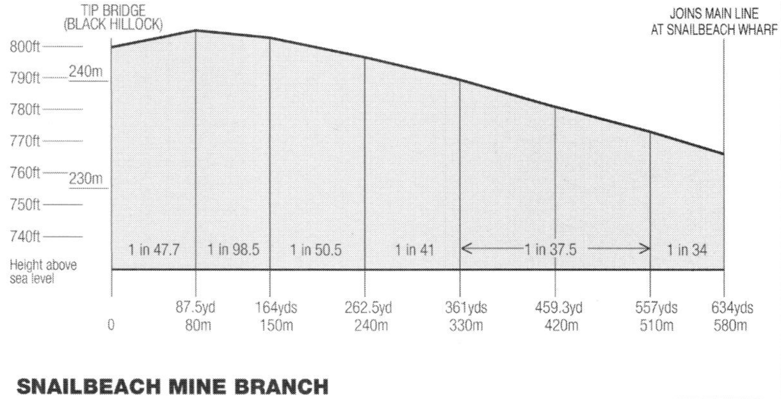

SNAILBEACH MINE BRANCH

SURVEYED BY MILES BEVAN, 1994

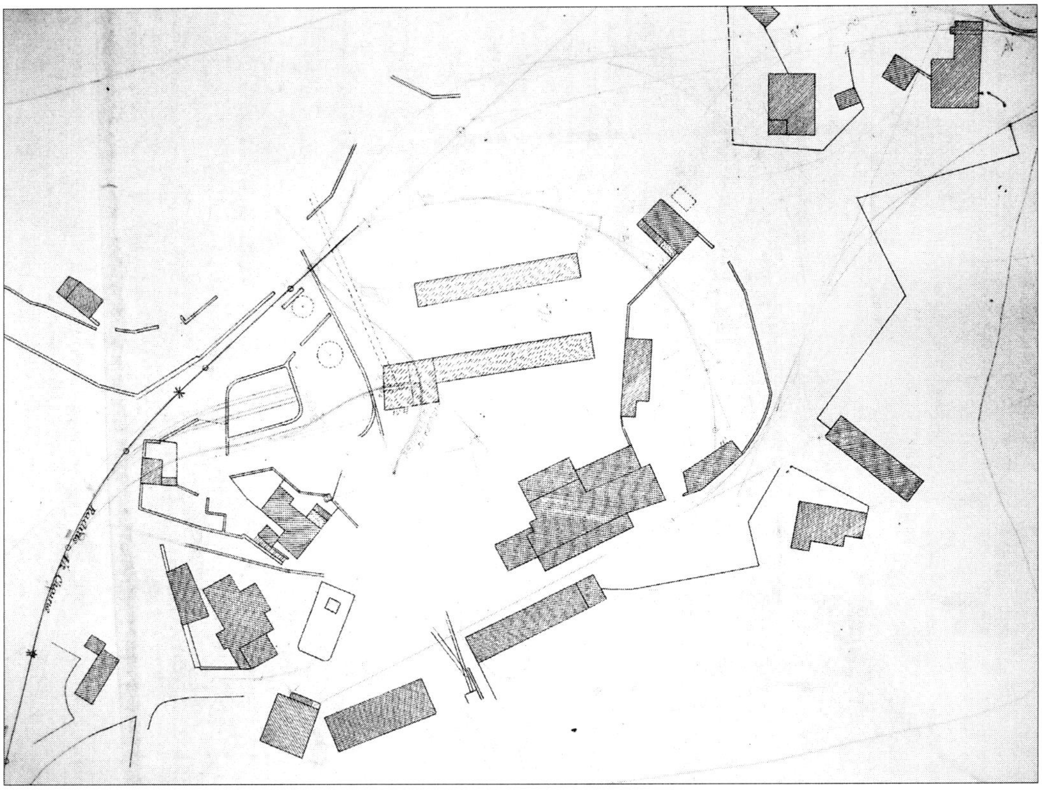

Working plan for the railway sidings at Snailbeach mine. Notably, it appears that only buildings complete with roofs are cross-hatched. The walling to be used for the locomotive shed was not. Note that the suggestion of a wagon turntable has been pencilled in lightly where the main siding was to cross the dressing floor tramway. A slip and 90 degree crossing were built instead. SHROPSHIRE ARCHIVES 4743/48

winch.[27] The coal thus raised fuelled the boilers for the mine pumping and winding engines.

Completion date of the line

Ample evidence has been given in Chapters Five and Six to show that Railway No.1 and the private sidings to the mine and smelt works were complete and working by late 1877. However, it was mentioned by Eric Tonks (2nd edition, p.21) that the chairs of the standard gauge siding at Pontesbury were dated 1881, leading to the suggestion that the junction was not brought into use until the SDR had been operating for some years. The present author admits to having fed this speculation in the journal of the 7mm Narrow Gauge Association (a railway modelling club), and also suggested that the sidings at Snailbeach mine were not complete until 1881. The latter idea arose from noting the cast iron counterbalance weights dated 1881 fitted to the point levers near the locomotive shed. Later relaying of track using chairs bought second hand could answer the first suggestion. The second might be answered by the need being found for more convenience (with safety) in the use of the points on the mine. Without a counterbalance weight or tedious locking and unlocking, derailments might easily have happened. All the sidings concerned are shown on the Ordnance Survey County Series 1st edition maps, which were surveyed in 1881. They must therefore have existed beforehand.

Ashbury & Co. weighted point lever dated 1881 at the Snailbeach locomotive shed sidings.

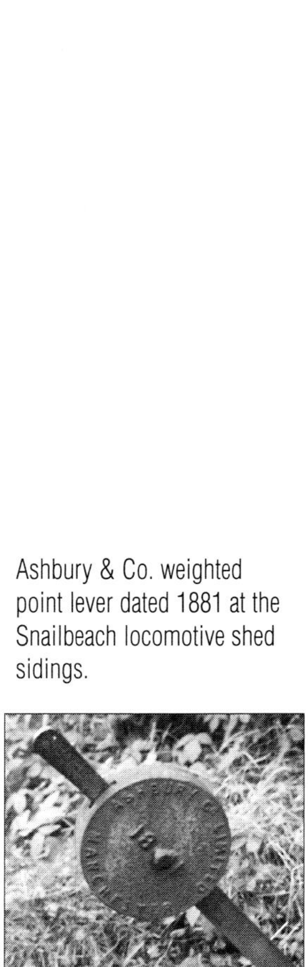

Permanent way maintenance from 1923

Apart from the initial specification for the track, little else is known about it in the Dennis era. In contrast a good deal of detail is available from 1923 onwards. Looking at the information shows the economical but responsible approach of the Stephens group to track repairs.[28]

Despite the change of ownership being almost a re-opening of the line, any work done on the track in 1923 appears to have been minimal, since no cost was entered for it in the Board of Trade Returns. No permanent way materials were bought other than for the replacement of 8 chains (176 yards) of fencing. In the next two years, a further 16 chains of fencing were renewed and sixty sleepers bought. These were carefully stacked between the tracks at Pontesbury, as can be seen in photographs taken in 1925. Permanent way maintenance for the two years cost £175. By the end of 1925, the coal siding had become disused, and the small amounts of coal now arriving via the Minsterley branch were being off-loaded into wagons standing on the old warehouse line.

More earnest work was planned for 1926.

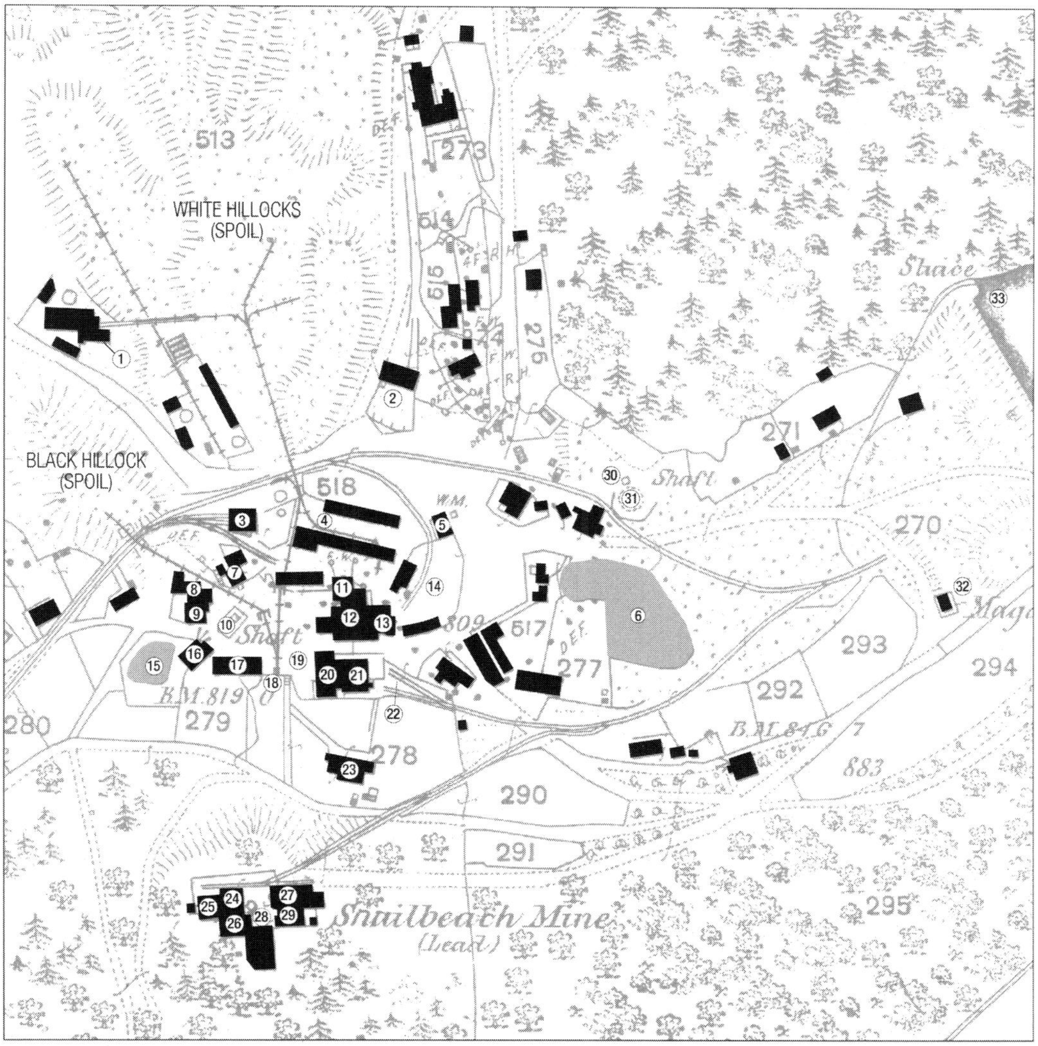

SNAILBEACH MINE

Based on first edition 25" OS plan

MINE TRAMWAYS AND TIP LOCATION

100 200 300 feet

100 m

1 CRUSHING ENGINE
2 ORE HOUSE
3 LOCOMOTIVE SHED
4 MAIN DRESSING FLOOR
5 OFFICE & WEIGH HOUSE
6 OLD POOL
7 OFFICES
8 SMITH'S SHOPS
9 OLD ENGINE HOUSE
10 CAGE WINDING HEAD GEAR (OLD SHAFT)
11 CRUSHING ENGINE
12 OLD CENTRAL BOILER HOUSE
13 CARPENTER'S SHOP
14 CARPENTER'S YARD
15 POOL
16 WINDING ENGINE
17 MINER'S DRY (BARRACKS)
18 EXIT FROM LANDING LEVEL
19 AIR RECEIVERS (3)
20 COMPRESSOR HOUSE
21 CENTRAL BOILER HOUSE
22 COAL YARD
23 COUNT HOUSE
24 WINCH ENGINE, INCLINE
25 BOILER HOUSE
26 CORNISH PUMPING HOUSE
27 BOILER HOUSE
28 NEW ENGINE SHAFT
29 WINDING ENGINE
30 BLACK TOM SHAFT
31 HORSE WHIM
32 MAGAZINE
33 RESERVOIR

The railway sidings, as completed in 1877, have been drawn by referring to photographs and OS maps. The 1881 O.S. missed two short sidings which were reinstated on the 1901 edition, and they appear here. The layout of mine tramways has been similarly reconstructed, but information is incomplete and their accuracy cannot be guaranteed. All of them in the central mine area appear to have been 2 feet 4 inches gauge.

The tramways were much the same around 1900 as they are drawn here, with the notable exception of the ore tip to the dressing floor, which was moved to tip directly into the crusher house via a curved trestle. With the Halvans Co. underground lease, 1912 on, work concentrated further east on the sett in the upper levels, there being dressing facilities at Black Tom shaft and in the Upper Works. Tramway materials were probably removed and relaid where required. The railway sidings seem to have stayed as built into the Stephens era, with the exception of the siding beside the locomotive shed, part of which was lifted when *Dennis* was scrapped.

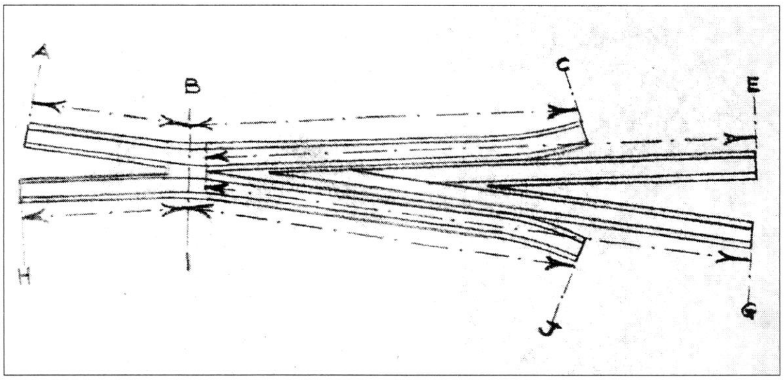

Diagram of crossings for railway points. The lettering is obviously with reference to a table of dimensions. A table exists which lists and so dimensions every crossing on the railway, but the lettering does not correspond.
COLONEL STEPHENS RAILWAY MUSEUM ARCHIVE

View south over the road bridge, Pontesbury. It took eighty years and a lease by Shropshire County Council to decide that hand rails were needed. In the foreground, notice a wrought iron rail has had its head partly sheared through with age and has been turned round and re-used. In the background can be seen the siding to the south tip. C. PEALLING

Looking towards Callow Hill at the Pontesbury bridge, the south tip siding branches to the right. C. PEALLING

This cost £170. Although small, traffic had all but doubled in three years, and the opening of the new road stone quarry at Callow Hill promised further increases. The track was taking much harder punishment. In January, Stephens told Will Jones that £90-worth of new rails had been bought, and exhorted him to organise the replacement of old ones which could then be sold to scrap dealers. The men were told to make sensible use of the best of the old rail, with Stephens instructing them:

> Where the old rails can be cut and used again for repairing places as shorter rails, I should prefer them to be used.
>
> The rails at Pontesbury Junction can be used for siding purposes and any good rails which are in sidings, for use in the main line.
>
> I do not want the shorter rails put in various places by themselves on main line.

The track was always well maintained, right to the end. In the Stephens era sleepers were always new, never second hand, and supplied by the Great Western Railway or local timber suppliers. At first they were pitch pine (see Appendix 2), later changing to oak. Dog spikes, bolts and the like were either shipped from the East Kent Railway (another line owned by Stephens) or bought from wholesalers locally. By turning the old wrought iron rails around, they were re-used after the heads had split through on the running side. Only when they had sheared on both sides were they scrapped. A full-time platelayer was always employed, and other staff helped with the track in their spare working hours. In 1941 it was reported by Graham Vincent that most of the rails were still the original iron rather than steel.[29] Wrought iron rails remained in service on the main line into the 1950s. Certainly in Austen's time, poles for fence repairs were sometimes cut from line side trees, and proper hedge laying could be as important as fence repair to maintaining the boundary with neighbouring farms.

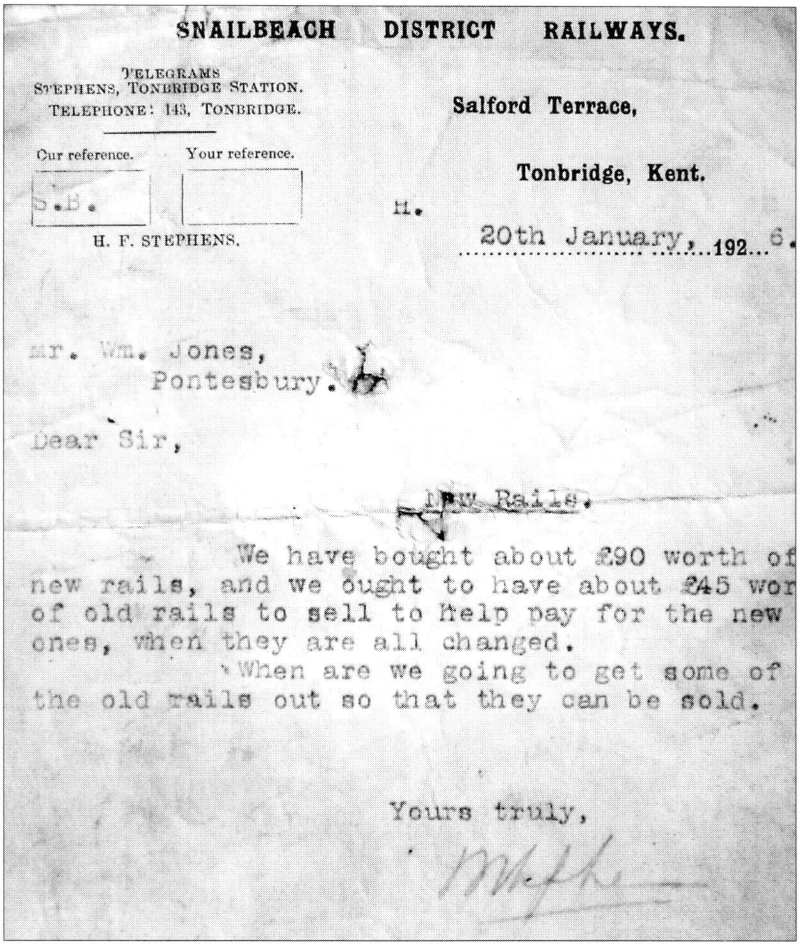

SNAILBEACH DISTRICT RAILWAYS.

TELEGRAMS STEPHENS, TONBRIDGE STATION.
TELEPHONE: 143, TONBRIDGE.

Salford Terrace,
Tonbridge, Kent.

Our reference. S.B. Your reference.
H. F. STEPHENS.

H.

20th January, 1926.

Mr. Wm. Jones,
Pontesbury.

Dear Sir,

New Rails.

We have bought about £90 worth of new rails, and we ought to have about £45 worth of old rails to sell to help pay for the new ones, when they are all changed.

When are we going to get some of the old rails out so that they can be sold.

Yours truly,

Memo from H. F. Stephens, 1926, encouraging Will Jones and the staff to pull out the unserviceable wrought iron rails and scrap them. COLONEL STEPHENS RAILWAY MUSEUM ARCHIVE

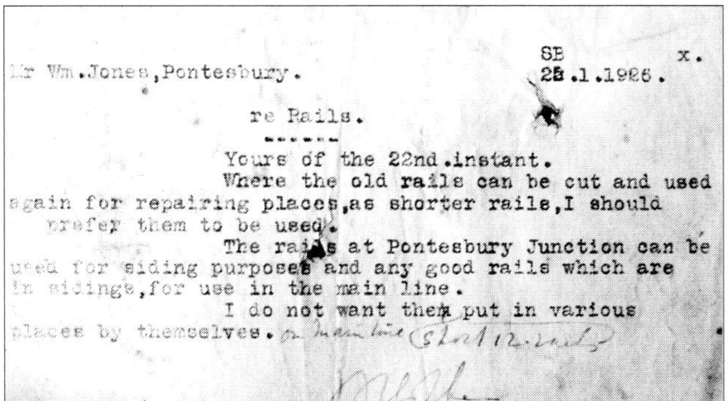

Mr Wm.Jones,Pontesbury. SB X. 25.1.1926.

re Rails.

Yours of the 22nd.instant.

Where the old rails can be cut and used again for repairing places,as shorter rails,I should prefer them to be used.

The rails at Pontesbury Junction can be used for siding purposes and any good rails which are in sidings,for use in the main line.

I do not want them put in various places by themselves. on main line

Memo from H. F. Stephens, 1926, detailing how to re-use the best of the old rail. COLONEL STEPHENS RAILWAY MUSEUM ARCHIVE

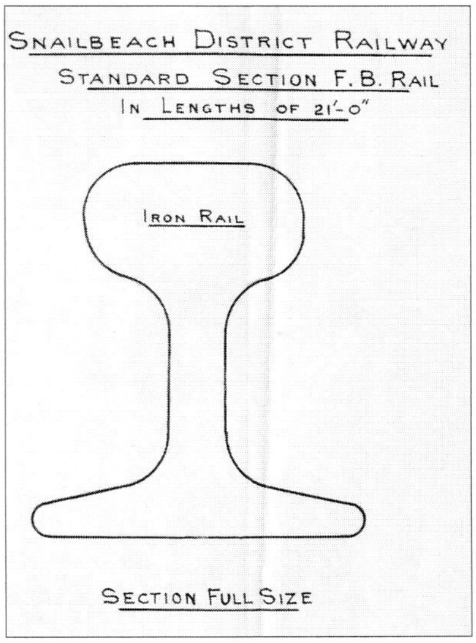

Cross-section of wrought iron rail. Joining such rails to modern steel replacements with a different profile required special measures. COLONEL STEPHENS RAILWAY MUSEUM ARCHIVE

The locomotive shed, 1935. *No.2* stands outside, with the remains of *Dennis* on the siding to the right. Note that the wooden draught stacks have still to be added. Comparison with the photograph c.1925 reveals a far higher pile of ash and clinker in front of *Dennis*; an accumulation of pipes, junk and debris to the left of the shed; and the dressing floor shed roof behind the loco shed in serious disrepair compared to a few holes ten years earlier. Other features of the mine shown include the crusher house on the far right, the carpenter's yard wall, and the ruinous mine office and weigh house. D. A. BREWER

CHAPTER ELEVEN
Railway buildings

In keeping with its economical construction budget, the railway was provided with the absolute minimum of buildings. The locomotive shed, a lime shed and wagon shelter, the weighbridge house, a platelayer's forge, a small office, and a large goods shed were built for the opening of the line. Of these, only the locomotive shed and weighbridge house survive.

The locomotive shed was on Snailbeach mine premises, and was included in a sub-lease from the mine company to the railway company for which rent was paid. In recent years it has been renovated, and is used as storage by the Shropshire Mines Trust. The weighbridge house at Snailbeach station has been absorbed into a larger building which has been extended to the north and west. The old goods doors have been replaced by French windows. It is now part of the domestic property known as 'Resting Hill'.

Nothing remains of the platelayer's forge near to the Callow Lane, and the office and goods shed which served Pontesbury sidings. At Snailbeach wharf, the stone foundations of the lime shed and wagon shelter can be found in the road vehicle goods yard. A couple of photographs were taken of the ruined forge in the 1950s, and the lime shed and wagon shelter are seen fleetingly in an old postcard. Apart from this, the most we know is contained in the Sale Particulars document and one contractor's drawing (see Appendixes 2 and 6), as nothing else survives of them other than outlines on maps.

Under H. F. Stephens' ownership of the SDR, a new wooden office was built at Pontesbury, close to the road. This position was convenient for receiving supplies delivered increasingly by road vehicle, and for connecting a telephone. Again, nothing remains of this building.

One other building should be mentioned for completeness. This is the old coal store in the Snailbeach wharf yard, built of wood and corrugated sheet, tenanted by Clarence Rowson, and erected probably in the 1920s (see Chapter Eight). It was not a 'railway' building as such, and more than likely he built the store to suit himself under the terms of a rental or lease agreement. In modified form this store survives today, and is now owned by the heirs of the late Douglas Challinor along with the Snailbeach wharf.

THE LOCOMOTIVE SHED

This building has been for many years the subject of many a railway enthusiast's camera. On such a simple railway there was little else of great interest to photograph, especially if the

Snailbeach wharf, rail-to-road transfer yard, in 1971. Doug Challinor had a lease of the wharf at the time, and he was making good use of the old garage and yard for his car breaking business. The grassy shelf next to Doug's shed was the site of the lower loop line, from which wagons were unloaded onto carts. The upper shelf on the left carried the main line. In the distance is Central Snailbeach mine.

No.2 outside Snailbeach locomotive shed with the remains of *Dennis* in 1935. Note how the ash pile has encroached onto the siding.
F. JONES, INDUSTRIAL LOCOMOTIVE SOCIETY, COURTESY A. C. BAKER

visit was on a day that there was no locomotive working. But it has never failed to fascinate, with its twin gabled roof, a trapezoidal ground plan, having two roads and standing in such an interesting location.

The accompanying scale drawing of the locomotive shed has been produced using historic photographs and information from pre-renovation field work.[1] A drawing showing the building as it was when in railway use would not be possible today without such information, because the renovation work has resulted in many renewed features being quite unlike they were. For instance, the window frames, roof vents and wicket gate differ, and modern weatherproofing methods have been used on the roof. The work has created a new building which is rather out of character with the old.

In part, the locomotive shed was originally a mine building, but its age and function are unknown. The layout drawing for the railway made in Henry Dennis's Engineering Office shows all the principal buildings and works in fair detail, where needed for planning the railway works.[2] On this drawing, sidings were planned, but they were not always completely drawn up to reflect how the railway was laid down. Such is the case with the locomotive depot. It shows a pair of tracks laid into an existing boundary wall (not a complete building with a roof: complete buildings are shown cross-hatched), which was part of a small network of structures in the area.

View showing south and east walls of locomotive shed, 24 May 1952. Window frame details are visible. On some panes there are extra vertical bars, where possibly two half-width panes are fitted. Also shown is the extended flue pipe from the hearth exiting the right hand ventilator slot. The siding to the left of the concrete platform is in thick undergrowth. The timber frame and pulley might have been part of a wagon tipping hoist.
F. W. SHUTTLEWORTH

LEFT: Archery slot ventilator above the south engine road, east wall, in 1985. The brick quoining and projecting purlins are visible.

ABOVE: Detail of smoke hoods. D. A. BREWER

It was clearly intended to use this walled area as the foundation of the locomotive shed. It is drawn with the trapezoidal shape, and is even shown curved in the north-west corner, where there is now a curved corner shelf inside the building. There is a similar curve shown in the south east corner. When converted into the locomotive shed, these two corners were built up outside to an apex, to give full structural support to the gables. Building material was local stone, some red brick and lime mortar.

The site was not ideal, because the long siding coming into the site was raised up slightly relative to the shed, promoting drainage towards the shed doors, but it was good enough. Efforts were made to plan sidings into the shed which did not involve the demolition of existing buildings, but some small structures including roofed ones shown on the plan were cleared away sufficiently to put in the shed roads as-built.

General building style and features

The roof covering was diagonally-laid boards on rafters surfaced with roofing felt. It was patched up heavily in later years, but began as a craftsman-like job. The truss and purlin timbers are marked with carefully chiselled Roman numerals at the joints. This suggests that the parts were cut to specific measurements, and could only fit one way.

It has been suggested that the roof was originally slated, but slates are not visible in the earliest photographs. A well-made boarded and felted construction could easily last decades, especially when the weatherproofing is properly maintained (just look at a veteran garden shed). Other sheds at the mine, erected by Dennis in the 1870s to house ore dressing machinery, had felt-covered roofs.[3]

Windows were simple non-opening wooden frames set into the stonework, topped by huge rough-sawn timber lintels. They were of uniform style, having two vertical panes, each puttied into an inner frame with a further surrounding frame and central divider. Some vertical panes appear to have been in narrow pairs, each with a further divider. In later years the surrounds were black and the inner frames white. Earlier colours are unknown. The bottom photograph on the opposite page shows them well. Two of the larger windows were put in the south facing wall, against which workbenches were built, to improve lighting. The odd sizes suggest salvage from earlier buildings. There were also roof windows.

The slot ventilators in the east wall and the dormer ventilators in the roof are original

Top hinge and post to the left hand door. Note the coach bolt through the post into the lintel. 1985. D. A. BREWER

This view shows the surroundings and numerous details of the locomotive shed. 24 May 1952.
F. W. SHUTTLEWORTH

features, the latter being visible in the panoramic photograph. Before the Great War, there were two draught stacks in the form of simple tubular metal chimneys, one over each shed road, but nothing is known of any smoke hoods inside (see panoramic photograph, pp.64-65) The Thomas view of c.1885 shows them clearly to be round and having flat disc caps mounted on legs. By the 1920s these chimneys had gone (see photo p.129). Wooden draught stacks with smoke hoods above the locomotives were added in the same positions after 1936. They were made by fastening pieces of board together, with louvres being trapped at an angle between the sides. The smoke hoods were sheet metal lined for safety against sparks. The external cowl tops were roughly formed from sheet metal and mounted on simple metal legs.

The locomotive doors are of noticeably unequal width, the left hand one having been either replaced or widened. To hang the wider door, the left hand door post was moved from a

The locomotive shed in 1971. The holes in the masonry to accept the extension timbers can be seen above the door lintel. Note the wicket gate with repaired panel underneath.

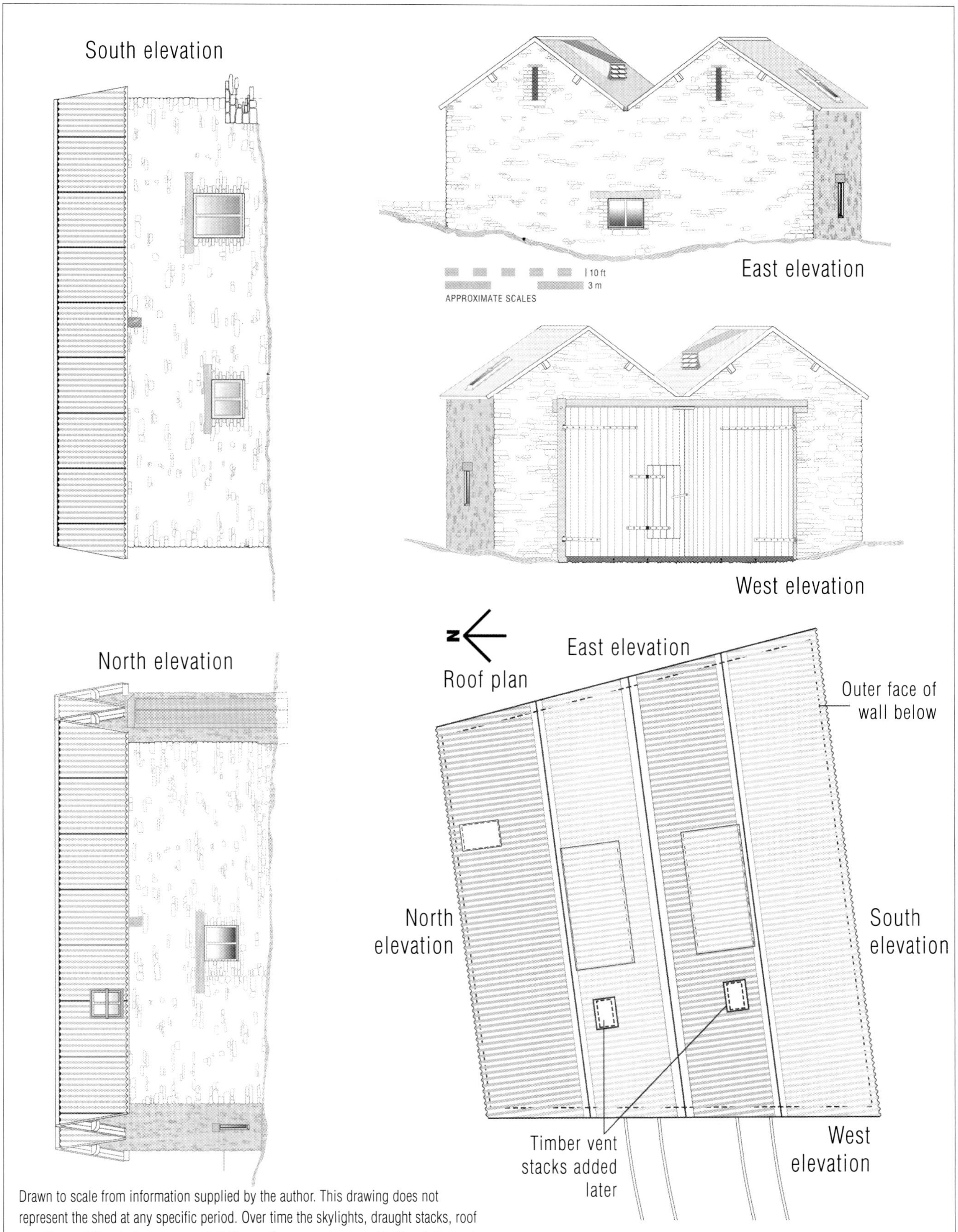

Drawn to scale from information supplied by the author. This drawing does not represent the shed at any specific period. Over time the skylights, draught stacks, roof covering and the window panes varied. These changes are apparent in the photographs.

Drawn to scale from information supplied by the author.

mirror-image position of the right hand one to the outside face of the wall, as shown in the drawing. The old post position was visible on the masonry. Moving the post gave five inches extra clearance, forced by the arrival of a wider locomotive than the railway was used to – either *Fernhill* in 1881 or *Dennis* in 1906. Overall dimensions of *Fernhill* are unknown, but *Dennis* was 7ft 6in wide across the roof. Note the wicket gate in the left hand door. This has a step-over, with the gate being set within the main door framing. The short boards at the bottom of the main door, under the wicket gate, were fixed to a frame member.

The extension roof was built in about 1923 to help get all four locomotives under cover. Some stones were removed above the doors to locate two of the horizontals of the framework. The corrugated roof stayed on until at least 1935, but a photograph by P. S. Halton shows it to have completely blown away by 1937-8.[4]

The weighbridge house

As well as being built of brick and slate, this had fashionable Victorian barge boards, ridge tiles, a clay chimney pot and tiling to the south end wall. All of these materials were supplied by Dennis's industrial concerns. Inside, a small cast iron fire grate and a flue were built into the south west corner.

The platelayer's forge

This was a brick and stone-built building with a slate roof. The only known photos show it in the 1950s in a ruinous condition with no roof. There was a traditionally constructed smith's hearth with a brick-built chimney in the north east corner, and a hand operated bellows. The door appears to have been facing the railway in the south west corner.

The lime shed and wagon shelter

A few details of these are given in the Sale Particulars, but these buildings can be seen partially and at a distance in an old postcard view (see page 97). An enlarged extract from it is reproduced here. In particular, the double gabled roof and the open timber uprights of the shelter over the short siding can be seen. The buildings stood abutted to each other. The shelter was to keep rain off any lime not unloaded from wagons. The arrangement of the two together can be ascertained from the given sizes and the stone foundations still extant, but apart from knowing that the shed had sliding doors, any other features can only be guessed at.

The weighbridge house, 1971. By this time the roof and gable wall tiles were a bit worse for wear.

An enlarged view of a post card showing Snailbeach wharf. The lime shed and the wagon shelter are visible but partly obscured by trees. The open post construction of the wagon shelter can be seen. COURTESY IRONBRIDGE GORGE MUSEUM TRUST

PONTESBURY GOODS SHED AND OFFICE (DENNIS ERA)

No photographs are known which show either of these buildings. Other than details given in the Sale Particulars and a contractor's plan of the goods shed, almost nothing is known of them, bar their appearance on Ordnance Survey County series maps and the Shrewsbury & Welshpool Railway plan of 1886. The office is not identified on either, but was located just north east of the coal siding cutting. The goods shed had a protective coating applied by the gas-tarring process in October 1878. This was extra work for junction man Samuel Jones, for which he was paid £2.[5]

PONTESBURY OFFICE (STEPHENS ERA)

This was a timber and roofing felt structure which appears in a few photographs. The most notable of them is by W. E. Hayward (see page 74), which shows it from the road vehicle side. The lean-to structure on the north side was probably an earth closet. Also shown in this photo is a station sign board, but unfortunately no lettering is visible.

There are a few points to be made about the buildings on this railway, concerning building materials, style and relative costliness.

The locomotive shed was mine company property, and was provided apparently at the least possible expense. It used materials and methods readily available at the mine, including rubble stone, timber and window frames, the latter perhaps being salvaged, and was built on existing foundation walls. The money for it did not come from the SDR construction budget, but instead from mine company funds. It was a sensible decision to have the locomotive shed provided by the mine company. The locomotives were best housed at the mine because they were most needed there (for shunting wagons), and it was where coal, water, labour and materials for locomotive maintenance were readily available.

All other original railway buildings were funded from the construction budget, and built with new materials which, provided the budget was not exceeded, could be transported in for the job. Such materials included roofing slate and Ruabon brick and tile products, which were all supplied from Henry Dennis's industrial interests in Denbighshire. Thomas Nuttall, one of Elias Griffith's partners in the SDR construction contract, was a timber merchant, and to Nuttall's interest (besides extensive amounts of temporary timber used in the railway works) some of the buildings included large proportions of timber.[6]

The most visible of the buildings – the Snailbeach weighbridge house – was the best appointed, and as described, was built in the Victorian style of the best and most fashionable materials. In its appearance it was part of the public face of the railway. In the case of the plate-layer's forge, there was a need for fire resistance, so brick and stone construction was essential.

No.3 hauling empty wagons in the cutting on 19 December 1941, approaching Woodhouse farm bridge. The photographer was banished to the wagons because the crew needed to concentrate on driving and sanding the rails. J. G. VINCENT

CHAPTER TWELVE
How the line was worked

Having described the engineering features of the railway, and its layout and equipment, the physical picture would be lifeless without some knowledge of how it was worked. Some of the following description includes first hand accounts, while the rest is a reconstruction of what was possible and likely with the equipment available.

The railway worked on a one-engine-in-steam basis, having no signalling system or passing place en route. There was no need for more. Journeys were kept to the minimum, although more than one locomotive working per day was made to Pontesbury if traffic demanded it.[1] Some traffic was gravity worked. Operation without the benefit of passing loops at the termini called for shunting methods which were not unique to this railway – gravity, towing by chain or steel rope, pinch-barring or even pushing by hand. The task of getting the locomotive from one end of a train to the other (for instance at Pontesbury, or at the mine) could be done more than one way. I have described some practical shunting movements, but to try to describe each and every one would be tedious and next to impossible. Interested readers can work out what else might have been practicable and safe. To follow the descriptions, the siding layouts in Chapter Ten should be referred to.

Locomotive workings to Pontesbury

Lead is a very dense material, and a typical 100 ton parcel of lead would have been carried to Pontesbury in comparatively few wagons. To get pig lead wagons and empty coal wagons out of the smelt works, locomotive haulage was needed, because the siding had an uphill gradient from the works to the junction. The rest of the trip was downhill. A mixture of shunting techniques would have been used on arrival at Pontesbury. Siding 5 was available to reverse the pig lead wagons into, thereby releasing the locomotive to reverse towards the bridge. The wagons might then have been pinch-barred back onto siding 1, coupled to the locomotive, and then braked down to the goods shed. Leaving the wagons there, with the first wagon parked by the indoor crane, the locomotive was free to do any other shunting. The junction man could tranship each wagon load, move the empty wagon clear through the far end of the shed, then bring the next full wagon into position. With the

19 December 1941. Derailing the end wagon. This was one means of discouraging joyriders from taking a wagon down to Pontesbury. Another method was to chain and padlock the wagon at the Pontesbury end to the track.
J. G. VINCENT

locomotive isolated, for instance on siding 1, any wagons brought behind it could have been parked south of the road bridge and gravity worked into the required sidings. Shunting chains were also available for towing wagons onto a diverging siding. The chains and two crow bars (which the author suspects were pinch-bars) are listed in Appendix 2. In later years there were steel ropes for such shunting.

LOCOMOTIVE *VERSUS* GRAVITY WORKING

Although it was possible to reduce some costs by gravity working, it could not in itself make the railway profitable. This always depended on having enough paying traffic to cover all costs. This is illustrated well by the changing fortunes of the SDR in the 1930s and 1940s (see Chapter Eight). The continuous down gradient from Snailbeach station to Pontesbury made gravity working convenient, but prevailed more in some eras than in others. This mainly depended on the nature and direction of the traffic. In the lead smelting era, by far the greatest traffic was coal in the up (locomotive hauled) direction. For coal to be brought regularly up to Snailbeach, regular locomotive workings down to Pontesbury were needed too. This in itself would have removed much need for gravity working, because most down wagons could have been sent with the locomotive. This can be contrasted with the Callow Hill quarry era. Then, road stone was the chief and eventually the sole traffic, was in the down direction and could be gravity worked.

PRACTICAL LIMITATIONS AND SAFETY ASPECTS OF GRAVITY WORKING

Aside from the gradients on the mine branch being rather steep, there was a small summit behind the village which ensured that all traffic from the mine was locomotive worked (see Chapter Ten). It was possible to operate by gravity directly out of Snailbeach station, because the track through the loading wharf was on a slight gradient.[2] Wagons being loaded on the line below Snailbeach station (including timber) could not have been worked safely by locomotive. This is because without extra shunting, the wagons would have been coupled below the locomotive, and runaways would have been inevitable if a coupling had slipped. The safest, easiest and most economical way of working such loads was by gravity. This cleared the way for the steam working if there was one, and got the goods delivered to Pontesbury on a particular day if there wasn't. Photographs show that two men could control and stop a gravity run of gross weight up to about 30 tons, and one man was safe with a handful of wagons. The usual routine was to pin down enough brakes to keep the speed down, while the brakesman kept control by stepping on the brake lever of his own wagon.[3]

There was obviously a good deal of risk for the men making the gravity runs, as can be imagined from the various descriptions below. But although there were derailments – some serious enough to put wagons out of action – there was little damage to life and limb.[4] The only reportable accident on the SDR involving a person *'injured while braking, spragging or chocking wheels'* happened in 1909 – it was not a fatality.[5] Local people have only recalled the accident to Will Jones' hand in 1919 (see Chapter Sixteen). Probably little if anything more serious happened after that, even with the intensity of the Callow Hill traffic. Clearly all the men working on the railway took great care of themselves.

GRAVITY-WORKED TRAFFIC

Loaded wagons were worked down the line by gravity from E. C. Gray's spar gravel works. Previous occupants of Gray's site (such as the South Shropshire Barytes & Lead Co. in the 1890s and F. W. Read) also used the railway.[6] Before the siding was put in at Gray's works late in the 1920s, empty wagons were left on the line for loading.[7] Gray was allowed to gravity work his own wagons at first, but after damage was reported to hopper wagons in 1926, H. F. Stephens forbade it.

The spoil heap siding north of Snailbeach (used by the Snailbeach mine companies, the Halvans Co. and later the Gravel Trading Co.) was suitable for gravity working, because the point faced the right way for Pontesbury. It was also easy to work full wagons down by steam because of this, and both methods were used.[8] Empty wagons were left at the siding by a combination of chain (later steel rope) shunting and pushing by the locomotive. Road stone traffic from Granham's Moor quarry was worked to Pontesbury by gravity from the tramway connection point.[9] Road stone traffic from Callow Hill was always worked down by gravity.

WORKING CALLOW HILL QUARRY TRAFFIC

After 1925, road stone again became the major traffic on the SDR, this time from Callow Hill quarry. In the 1930s, the Halvans Co. completely

19 December 1941. Switching temporary rails for the siding at Callow Hill quarry. Graham Vincent noted: 'A primitive substitute for points – slewing the track. This was done twice a day, 3 days a week for six weeks.' After this a normal point was installed. From back to front, the men pictured are the fitter-driver (either Gostlow or Dodd), his assistant on the locomotive, and probably one of the quarrymen. Some quarrymen were assigned to transport work. At the rear of the picture, the empty hopper wagons are waiting to be pulled into the siding, while the steel shunting rope can be seen laid out to the left of the track. The locomotive has pulled forward and is behind the photographer.
J. G. VINCENT

changed to motor transport of barytes from the Upper Works. Steam driven machinery gave way to petrol and oil engines too. Coal traffic reduced considerably after 1932, probably affected by the death of coal merchant Clarence Rowson, and as a result of all these factors freight other than stone carried in the 1930s became insignificant. Paying traffic was almost without exception in the down direction, and mostly let down by gravity. Most up traffic was empty wagons, the majority being for Callow Hill. But the demand for wagons was increasing. In autumn 1927 a locomotive was steamed four days a week, and on five days a week in January 1934. Over this period, traffic more than doubled, and the demand for wagons at Callow Hill forced the re-supply of empties on almost every working day.

After 1940, Callow Hill stone was the only traffic, with the exception of a few tons of fuel a year for an unknown customer. Traffic dropped to 1920s levels, and the locomotive was steamed only three times a week. The locomotive ran light to Pontesbury. The empty wagon order for the next two days was marshalled and hauled up the line, this being all the locomotive was needed for. A memo from Austen to Bill Jones in 1939 instructed the men to avoid steam workings during the quarrymen's holiday week. A week's supply of wagons for the spar workings was to be hauled up to Snailbeach in advance.

Rope shunting at Callow Hill, 19 December 1941. The quarryman is riding the wagons and holds the tow hook in place with his foot. Although a poor quality photograph, it is included because it is a unique record. J. G. VINCENT

Callow Hill gravity runs

Wagons were batched, with each carrying either ½in or ¾in graded stone chips and sent down separately. But just as importantly, the number of wagons in the batch, or 'run', had to be limited for safety reasons. The men had to be sure that the run could be stopped at Pontesbury. Points accessing sidings at Pontesbury were set and padlocked to the correct position as the previous up train passed through. Often the run was eight wagons long, and there would be two men aboard. Sometimes a flat wagon would be coupled to the back of the run, and the men would ride on this rather than cling onto the sides or between the hoppers. Derailments were not unknown, and it was far easier to escape injury if you were riding on a flat.

When night shifts were worked at the quarry in the late 1930s, runs were worked down early in the morning when it had turned light enough to see. These runs would be up to a maximum of ten wagons, and very difficult to stop because the rails were so wet and greasy with the dew fall. So much braking was needed that it was easiest to

Gravity run in the cutting, 19 December 1941. On the far bank is a slightly neglected mile post. J. G. VINCENT

11 August 1954. Gravity run - brakesman unknown. The rear wagon shows the precarious perch for a brakesman, and position of the brake lever for his outstretched foot. In this instance the brake on the rear wagon is pinned down. A. J. PRATT

jam the wheels. When the run was under way, the men worked along the wagons, clinging on and at the same time bending down to 'sprag' one pair of wheels at a time with a piece of iron bar. This was repeated until the run was felt to be going at a stoppable speed.

Road stone transhipment at Pontesbury

Granham's Moor stone was dropped from hoppers into standard gauge wagons. Hayward's Quarries initially distributed the stone by rail, for which it had standard gauge private owner wagons. With the increasing use of road lorries, there was a move to road distribution, for which the sidings were not equipped (see Chapter Eight). Before the mid-1930s, delivery onto lorries at Pontesbury was primitive to say the least. Uncoated chips of one grade would be emptied north of the road, next to the yard office, while the other run would be stopped just south of the bridge. The embankment south of the road was specially lined with paving slabs.

No.4 and a train of empty wagons above the south 'chutes' at Pontesbury about 1938. The driver is probably Jack Owen. Note the 'fences' for segregating the stone by size. H. F. G. DALSTON

Pontesbury road under bridge, 1933. The 'chutes' for road stone can be seen as pale, bare banks beside the railway, in the left and right background.
W. E. HAYWARD

These two points were known as the 'chutes' in railway record books. Old box wagons were used. The chippings were shovelled out of the wagons through the side doors and thrown down the bank. Using hopper wagons which dumped the chippings straight onto the track was little short of impossible. Once unloaded, this mass of chippings had to be shovelled off the bank and onto the waiting lorries. Most of the work was done by the quarrymen who had brought the run down, but the lorry drivers were expected to get it done, help or no help. They always handed round cigarettes to make sure there were plenty of willing hands. The loaded lorries were passed over a weighbridge on the south side of the road. Even in the 1930s, with for example Sentinel steam lorries which look small today, each lorry load was well over twelve tons.

An idea that worked for a couple of years was to load the lorries under the bridge. Some decking was taken up from between the rails, and the wagons were stopped here. The stone was shot through the rails into a lorry waiting below. But there were problems. The long drop from the wagon meant that some chips always bounced out onto the road. Motorists complained about these stray chips being thrown up and hitting their windscreens. It also created a cloud of dust.

A loaded run waiting at the lorry chute. A couple of empty wagons can be seen which have been pushed away. Notice the retaining wall was robbed to build the new loading bay, which is already in disrepair after less than twenty years.
F. W. SHUTTLEWORTH

At this time, two Sentinel steam lorries were used, and the steam from the waiting lorry combined with the dust made such a fog that no-one on the road could see. This together with parking under the bridge made it very dangerous. It was labour-saving, but it had to stop.

In the 1930s Shropshire County Council was renting the Pontesbury railway yard, and a tarring plant for pre-coating the stone was installed with a short passing loop siding laid in to serve it (see photograph page 75). The plant was owned by Thomas Swann but operated by the Council. These stone chips were at last unloaded the easy way, by opening the trap doors of hopper wagons directly onto a purpose-made chute. Full wagons were run by gravity into a shelter over the chute, and empties hand shunted out. From the chute, the chips were picked up by an elevator which carried them up to the tarring plant. Chute delivery of uncoated chips to lorries was in place by September 1938 (shown by a dated photograph by R. E. Tustin, not included in this book). For this, the tarring plant loop was extended southwards and a loading bay built. The chips were dropped into an underslung hopper with a lever-operated shutter, which could be used to help spread the load throughout the lorry. Again, empty wagons were pushed away to the west siding.[10]

Such progress did not get rid of all the shovelling! The smallest screened stone, known as '1/8th-to-dust', was useless at that time.[11] Now that the useful stone was taken directly away, space on the banks at Pontesbury, both north and south of the road, was used for stockpiling this waste material. In the main yard, this stuff was shovelled from the track as before, but south of the road a siding was laid in the tractor-hauled era which allowed it to be dropped straight on the heap.

The lorry loading bay at Pontesbury with full wagons waiting above.
COLLECTION D. A. BREWER.

3 September 1955. From lorry loading bay looking south, showing the office, the weighbridge, the south siding and stockpile.
A. J. PRATT

Snailbeach wharf in 1941. with Nurse Hand's garage in the road yard. In the foreground the mine branch begins its steep climb to skirt round the village. R. K. COPE

SNAILBEACH STATION

Having road and rail weighbridges at Snailbeach permitted two-way traffic, and there was some of each. Ore from Tankerville, Bog, Pennerley and other mines was loaded onto the railway. So was local timber (see Chapter Seven). The Granham's Moor Quarry Co. made some deliveries onto road transport using its side tipping wagons.[12] Coal for the Tankerville, Bog and Pennerley mines was also delivered to road transport here.[13] The locomotive crew would have probably helped the weighbridge man by shunting loaded trains over the weighbridge, one wagon at a time.

THE MINE AND SMELT WORKS

Access to the coal yard and the incline via a head shunt meant that coal was hauled to that point, then propelled to its final destination. By switching wagons between the two sidings joined by the head shunt, a train which was longer than the head shunt could hold could be delivered to the boiler houses. At the same head shunt, a wall was built which allowed carts to tip minerals into railway wagons. A small heap of spilt mineral at the base of this wall, next to the track bed, was seen in the 1990s. The Halvans Co. hired carters to haul minerals from Perkin's level to the railway.[14] The head shunt was one place where the transhipment took place. The tipping carts, which had a removable pin to allow the body to tip backwards, are recalled today.

The down gradient from the smelt works junction to the works was helpful to the train crew bringing coal from Pontesbury. It was possible to release the locomotive by leaving it on the east siding, then running the wagons into the works by gravity.

ADDITIONAL MANNING; HEAVY AND BULKY ITEMS

The task of carrying metals, ores and coal to and from the mine and smelt works was the reason for existence and the life blood of the railway. But the railway company did not provide the labour for these jobs other than the locomotive crew, the junction man, and later a labourer. Bulk loading, unloading and the movement of stocks between Pontesbury, the mine and the smelt works was let to companies of miners in the customary manner of running the mine. This included shovelling in and out of railway

The slag house at Snailbeach smelt works on 28 February 2012, now cleared of a century of undergrowth and with roof mounted ventilators repaired. The gap in the wall close to the building was the railway entrance for coal, coke and other stores.

wagons and tipping from barrows, lifting and carrying, especially of pig lead, and moving wagons both empty and full between the dressing floors and the main siding. There was also a Loco Shunter employed at the mine (only known through a mine fatality – see Chapter Sixteen), who helped the locomotive crew with assembling trains and with coupling and uncoupling. All such work was considered to be a mine or smelt works cost, rather than a railway cost. The little that is known about the jobs done by the permanent railway staff is described above and in Chapter Fifteen. But for the junction man, cranes for very heavy or bulky items such as pig lead, timber and machinery were installed at Snailbeach and Pontesbury. These and the other labour saving devices, such as the coal wagon tippler and the construction of the sidings at various levels for gravity unloading, all reduced cost for the railway company by allowing quick and effective operation with a remarkably small number of staff.

Carriage of explosives

The SDR carried gunpowder as well as other mining stores. For a railway to carry explosives, it

28 February 2012. Site of smelt works siding, looking towards the junction and descending to the east side of the furnace hall, Snailbeach. This siding brought lead ore in from the mine, and it was a considerable height above the furnace hall floor. This gave the advantage of gravity for unloading ore and feeding the furnaces. In the distance, all traces of the railway branch siding embankments have been completely ploughed out from the neighbouring field.

Pontesbury sidings c.1925, showing the interchange pier and coal siding, with the derrick and a heap of timber in the background. This timber was being cut and shipped out by timber merchant Jabez Barker. In the foreground, the standard gauge siding is walled on both sides. COURTESY D. A. BREWER

needed to operate under byelaws specified under the Explosives Act 1875. All kinds of explosives were categorised by the Act, and provisions for safe carriage and handling defined. A draft of the standard byelaws concerning carriage on railways, probably for printing a public notice, survives in SDR records.[15] An identical printed and signed byelaws poster for the Talyllyn Railway is in The National Archives.[16]

Although the regulations specified the construction of special explosives carriages, at that time the railway had no need for one. There were safe ways of carrying gunpowder in ordinary rolling stock. The SDR had an explosives van, but only some years after the line was opened, and much later than when this Act came into force.

Article 6 of the byelaws stated that:

> Gunpowder, or any explosive made with gunpowder, ... if packed in metallic cylinders of a pattern approved by the Company, and similar in construction and security to those used by Government for the conveyance of small quantities of gunpowder by railway, may be conveyed along with ordinary goods traffic in a carriage not containing any article or substance liable to cause or communicate fire or explosion.

Neither were such vehicles required to have any special label attached to them concerning gunpowder carried this way, but the position of the vehicle in the train was a concern. Wagon labels were mandatory for all other explosives. Article 9 stated that:

> On each side of every carriage containing any explosive there shall be affixed in conspicuous characters, by means of a securely attached label or otherwise, the word 'explosive' or the name of the explosive with the word 'explosive' except when containing gunpowder or ammunition packed in metallic cylinders as provided for in the 6th of these byelaws; and every carriage containing explosive shall be placed as far as practicable from the engine attached to the train.

Locomotive coal, water and stores

While the railway was directly associated with the mining company, coal and stores for the railway were provided at the mine, and bought on account.[17] The locomotive shed was rented from the mine company and locomotives took on coal at the coal yard. Under the Halvans Co. lease, the locomotive shed with its water supply, the locomotive coal yard and the carpenter's yard were sub-let to the SDRCo. This sub-lease was still in force under Charles Moore's ownership of the Halvans Co. (See Chapter Eight.) The earlier rental of the shed probably included at least the locomotive coal yard and water supply in a similar fashion. Other watering points are mentioned in Chapter Ten.

It is not known when deliveries to the coal yard at the mine stopped. Hauling the locomotive coal to Snailbeach effectively increased its cost. From about 1926, locomotives took on coal on the west siding at Pontesbury. Even after this, for instance in 1927, some wagons of coal for locomotives were taken up the line. From 1934 coaling at Pontesbury is the only arrangement remembered. Some later photographs show single open wagons presumably being taken to Snailbeach, but they might have been carrying goods or coal for customers, or even stores for the locomotive shed.

From 1923, coal and all manner of stores were ordered by the Tonbridge office and delivered at Pontesbury. Local suppliers were used wherever possible. The fitter-drivers were hard pressed by the Tonbridge office to economise on coal. They did their best, but surviving correspondence shows that significant amounts went missing. Many local people could not afford to buy coal. Many gathered firewood in the woods, and others were tempted into stealing coal from the railway. With one of the locomotives always standing outside the shed overnight, it was easy to lift a large lump or two from the coal bunker under cover of darkness. The coal wagons at Pontesbury were even more vulnerable. The sidings were some distance from any houses and shielded from view by trees, so it would have been quite easy to visit the wagons after dark with a couple of buckets and a shovel. With the railwaymen being under pressure, the pilfering was not ignored. David Jones wrapped barbed wire over the locomotive bunker and put a rat trap inside. No-one has mentioned if any guilty party was ever discovered.[18]

The pressure for fuel economy was felt in other ways. Bill Austen made it known that the usual five hundredweight of coal burnt on a locomotive working day was excessive. Although the locomotives bought by Stephens were generally adequate, working of the empties up to Callow Hill was not always easy. The sand boxes saw good use in avoiding wheel slip and having to divide the train.[19] The latter would have meant an extra journey and increased coal consumption to answer for. Locomotive sand was bought regularly.

The Minsterley branch line service

A Joint Railways operating timetable dating from the 1930s and issued to staff shows the goods service available at Pontesbury Junction. It is unlikely to have changed much, if at all, during the life of the SDR.

On Mondays to Fridays, a goods train from Coleham to Minsterley called at 'Snailbeach Junction' at 12.23 p.m. with seven minutes allowed for shunting. It then departed at 12.30 and arrived at Minsterley at 12.40. This brought in coal and other supplies and empty wagons for minerals. On Saturdays, a similar train stopping at Pontesbury (departing 10.15 a.m.) called as required at the junction and arrived at Minsterley at 10.30.

In the down direction, a weekday goods train from Minsterley, departing 2.40 p.m. to Pontesbury only, called at 'Snailbeach Junction' at 3.01 p.m. There were nine minutes allowed for shunting, departing at 3.10 p.m. and arriving at Pontesbury at 3.15 p.m. Any through goods from Pontesbury was picked up by a later fast goods train from Minsterley.

An empty wagon journey in the cutting on 19 December 1941, heading for Callow Hill quarry. Junction man Will Jones is helping on the footplate by operating the sanding gear. As described in the text, this was important in helping to avoid expensive second journeys.
J. G. VINCENT

No.3 shunting empty wagons on the west siding at Pontesbury in about 1940. J. G. VINCENT

CHAPTER THIRTEEN

Motive power

All the SDR's locomotives were steam-powered. From the opening of the line in July 1877 until 1881, the sole motive power was the locomotive *Belmont*. In 1881 the second locomotive *Fernhill* was bought. This brought the railway several benefits. As described below, using *Fernhill* increased the uphill haulage capacity while possibly making operation of the line safer; and having two locomotives meant that there was a spare to cover for routine maintenance or breakdown. With increased traffic again in prospect, a new locomotive *Dennis* with the power of *Fernhill* was bought in 1906, to replace one of the earlier ones. High costs during the Great War left the company unable to buy any replacement locomotives, and it was forced to hire horses between 1920 and 1922. From late-1922 under new ownership, three locomotives were bought which gave plenty of spare capacity. Traffic requirements by then were less stringent than in the lead mining era, and these replacements were generally less powerful than their predecessors.

1877 – 1922

Belmont

The railway's first locomotive was named after the seat of the senior member of the Lovett family, Belmont Hall (known today as Henlle), near Gobowen. It was built by Henry Hughes & Co. of Loughborough, and was a 0-4-2ST weighing around 13 tons in working order. A cab with the rear open above waist level was fitted. Its first known owner was Ifton-Rhyn Collieries Ltd, a company incorporated in 1873 to mine Lord Edwin Hill-Trevor's land east of Chirk and the Llangollen branch of the Shropshire Union Canal. The village of Ifton is in north west Shropshire, close to the Welsh border. A 2ft 4in gauge tramway was constructed, intended to tranship coal onto the Shropshire Union Canal south of the colliery. A lease was made in 1874 to mine the coal under the Henlle Hall estate of T. H. Lovett, which was also to the south, allowing the line to be extended to the Great Western Railway at Gobowen. The eventual intention was to make the entire railway standard gauge. Part of the narrow gauge railway was working by 1874, over which sand and clay were hauled to the colliery's brick works by the Hughes locomotive, then named *Salome*. But in 1876 the Ifton-Rhyn company was in liquidation and *Salome* was laid up awaiting sale. The railway was never finished, but some track bed was re-used when W. Y. Craig & Sons Ltd re-opened Ifton colliery and completed their standard gauge railway to Weston Rhyn in 1921.

The SDR was already under construction when Henry Dennis heard of the Ifton locomotive being for sale. On 15 February 1877, Elias Griffith and a team of labourers, joiners and carters recovered the locomotive from the Ifton railway and hauled it to the Cambrian Railways' works at Oswestry. Here the locomotive had a general overhaul and was repainted with a red lead finish. New name plates do not seem to have been fitted – none were listed on the Cambrian Railways' bill, while the value of the old ones was deducted from the bill. The Cambrian Railways delivered *Belmont* carriage paid to Pontesbury station by rail at the end of June. Elias Griffith arranged carting for the transfer to the SDR, since Pontesbury Junction was incomplete.[1]

In some ways the locomotive was ideal – for a start it was to the same track gauge as the SDR. The total cost of putting *Belmont* into service on the SDR was £159 13s 2d, including the purchase price of £50.[2] Capital was a problem for the directors, and this low cost was helpful. When the offer of *Salome* came along, Dennis probably had no choice but to buy it. There is no way of knowing if Dennis expected the locomotive to perform as he wished, but in all likelihood it did not.

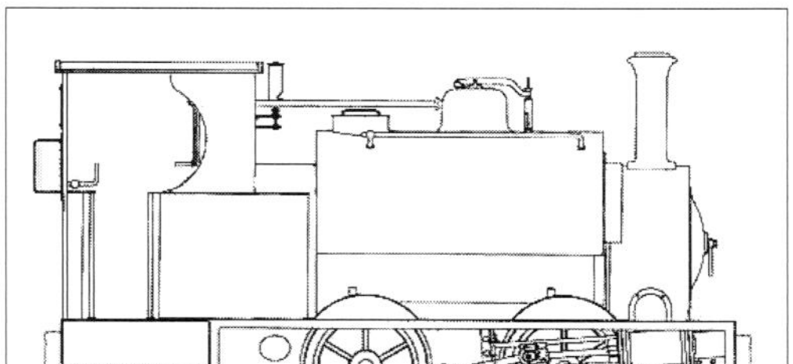

Conjectural drawing of *Belmont* c.1900 based on photographic evidence. NIGEL R. TAYLOR

Fernhill on the long siding with crew and miners at Snailbeach mine, date unknown. This is the only known photograph of this locomotive, but it looks very new, so it could date from the 1880s. The crew members have not been identified. Timber is being hauled on a flat wagon, with other wagons behind. In the background, there is a row of coal wagons on the coal yard siding.
NATIONAL RAILWAY MUSEUM/SCIENCE & SOCIETY PICTURE LIBRARY

Belmont was not, as has sometimes been assumed and stated as fact, the same as Hughes locomotives built for the Corris Railway (CR), which served slate quarries north of Machynlleth in Wales. Research by Nigel Taylor and reaffirmed by the author proves this. All known Hughes saddle tank locomotives of the 1860s and 1870s had a strong family resemblance, much as locomotives by other builders did, but this is where any similarity between the SDR and CR locomotives ended. Henry Hughes & Co. was no more confined to building one type or class of locomotive than other builders were. Research by Roy C. Link further shows that the CR locomotives were based on the Hughes street tramway locomotive, a design which post-dates *Salome*. Hughes & Co. built its first prototype tramway locomotive in 1876, two years or so after the Ifton-Rhyn railway was first operating. Following this debut, the tramway locomotive only appeared in any numbers in mid-1877, when they were put to work on the Vale of Clyde Tramways, Glasgow. It was impossible for Hughes to build a locomotive in 1874 using a design which did not exist until 1876.[3] The CR locomotives were built in 1878 as a 0-4-0ST type with a 'plantation'-style cab, which was open all around above waist level and the roof supported on four thin pillars. They had a pair of trailing wheels added later. *Salome* or *Belmont* was built as a 0-4-2ST type, with a cab having only the rear open above waist level. As well as being rather different in general outline from the CR type, *Belmont* was also some two or three tons heavier, and had larger driving wheels and cylinders. It was a larger and more powerful locomotive. Nigel Taylor's side view 'impression' drawing is based on a photograph of *Belmont* c.1903, which was seen many years ago but has more recently proved impossible to locate.

Fernhill

The railway appears to have had some difficulty operating with *Belmont* alone. The SDR directors reported on 16 February 1880: *'In view of the present and prospective condition of the Traffic of the Line, and of the possibilities of accidents, your Directors consider that the time has arrived when additional Locomotive power should be acquired, and they accordingly recommend the immediate purchase of a second Engine ...'* *Fernhill* was bought new for £900 through the sales agency Lennox, Lange & Co., which had offices in London, Manchester and Glasgow. It was a 0-6-0ST design by Barclays & Co. of Kilmarnock, and named *Fernhill* after another Lovett family seat, near Whittington,

The W. G. Bagnall works photograph of *Dennis*.
COURTESY D. A. BREWER

Shropshire. On 23 August 1881 the directors reported the new locomotive received and working. It carried works plates showing the builder as Lennox, Lange rather than Barclays & Co., which was often the practice for agency sales of locomotives. Weighing about 19 tons in working order, it had much in common with the more usual Barclays 0-4-0ST locomotives of that period, which were more often built for track gauges from 3ft 6in up to standard gauge. In this case, for such a narrow gauge, Barclay put the wheels inside the main frame, and spread the weight by using a long frame and adding a third axle under the cab and behind the firebox. This arrangement allowed the use of a conventional design of boiler and firebox, with no compromises on boiler tube surface or firebox grate area and giving plenty of steam. The long wheelbase dictated flangeless centre driving wheels, to ease progress of the locomotive through curves of the tightest radius. Water tank capacity was 570 gallons. The design pushed the axle loading to a maximum, on the wrought iron rails specified at about 40 lb per yard.[4] The livery is unknown.

Dennis

In 1905, the Granham's Moor Quarry Co. Ltd of Shrewsbury was forced to use the SDR, rather than transport stone by road from its quarry close to Habberley, on the east side of the Stiperstones. New investment in the railway to deal with the expected extra traffic included taking a mortgage to buy a new locomotive (see Chapter Seven). This was *Dennis,* a 0-6-0T, ordered in December 1905 by Dennis & Son and built by W. G. Bagnall Ltd of the Castle Engine Works, Stafford to 2ft 4in gauge. It was their works number 1797 and despatched to Pontesbury in February 1906.[5]

Dennis replaced one of the earlier machines, but which of them was withdrawn is unknown.[6] The withdrawn locomotive was then broken up on site and the boiler used by the Halvans Co.[7] The trade magazine *The Locomotive* of 5 March 1912 said that *Dennis* '*... was designed to take a train of 100 tons behind it and pass round 300 ft. curves on the main line and 112 ft radius curves in the sidings ...*' and some dimensions were listed. The centre driving wheels were flangeless. Water capacity was 500 gallons and steam brakes were fitted.

The surviving nameplate of *Dennis*.
COLONEL STEPHENS RAILWAY MUSEUM

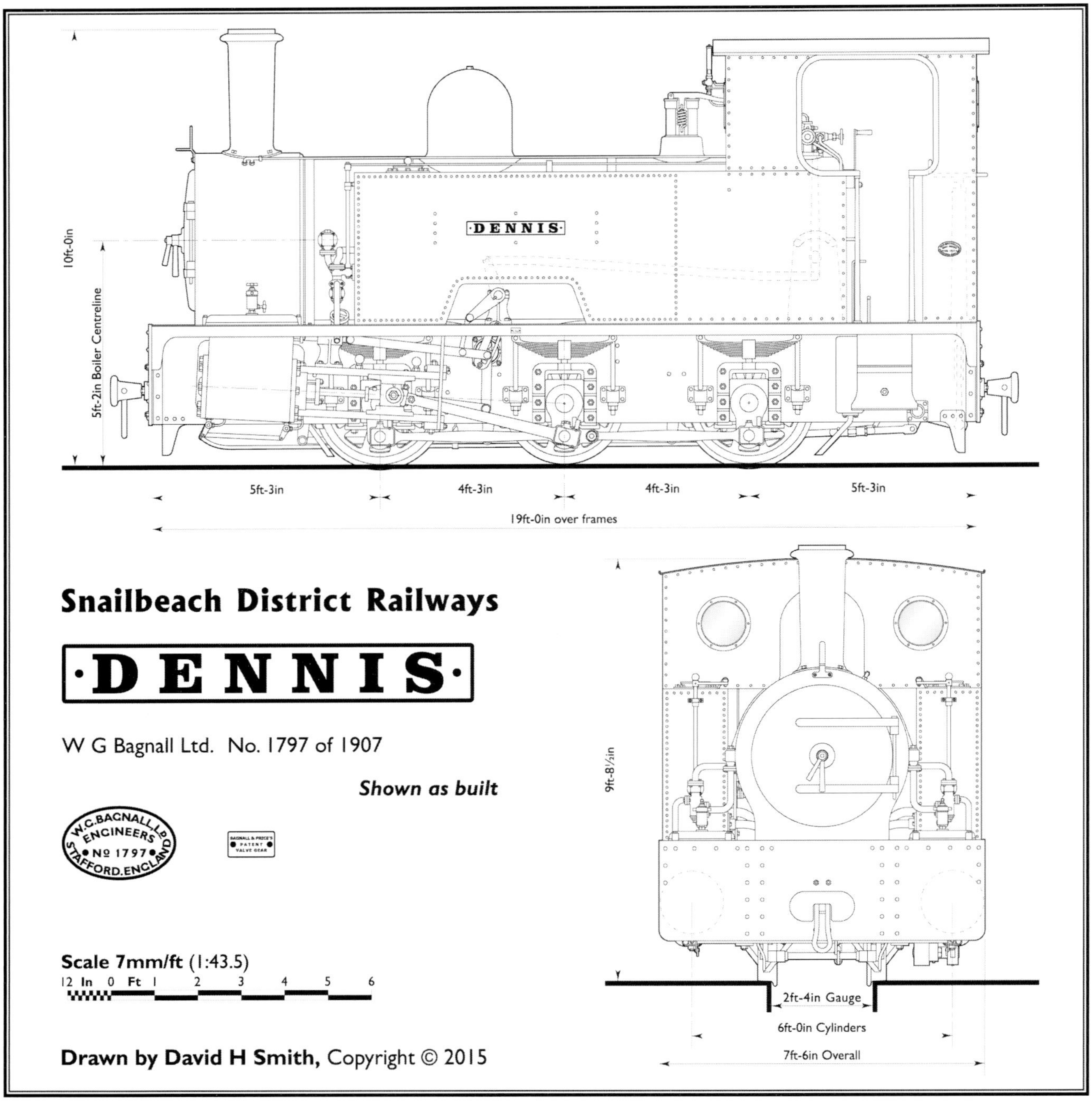

Dennis was fitted with brass name plates and painted larch green with red and yellow lining.[8]

A major overhaul of *Dennis* was made between November 1910 and January 1911. Parts fitted new included 25 boiler tubes of either steel or 'charcoal iron' bought from a Wellington manufacturer; a brake screw bought from W. G. Bagnall; and a set of fire bars and brake blocks (with a set of the latter put by as spares) from E. Davies of Shrewsbury.[9] To cover the time *Dennis* was out of service, there was still a spare older locomotive available. Drawings and sketches made after 1922 show that the firebox of *Dennis* had half side patches of copper added at some time.[10] Major overhaul intervals of five or six years were to be expected, and the company secretary's letter book indicates no further overhauls between 1910 and 1920 – it notes only the problems resulting from unreliable motive power – so it seems certain that this patching was done as part of the 1910 overhaul.

The second of the two earlier locomotives was withdrawn in 1913, leaving the railway without a spare to cover repairs and maintenance. Total reliance on *Dennis* brought problems for the company and its customers when major overhauls were needed. By 1916 *Dennis* was unreliable, and even out of service at times. The SDRCo had neither the cash nor sufficient prospects of traffic to allow it to buy another locomotive, although it appears to have spent some time looking for one. By 1920 *Dennis* was permanently out of use, and when required horses were hired from local farmers, who often traded as hauliers (see Chapter Seven).

A practical comparison of *Belmont, Fernhill* and *Dennis*

Fernhill and *Dennis* were rather larger than might be expected on a light railway of such a narrow gauge, but a look at the traffic and operating methods of the railway shows that acquiring locomotives of this size was a matter of policy. Experience drove the directors to look for heavier, more powerful locomotives. A clue to this can be found in the directors' report quoted earlier, where one reason for the decision to buy *Fernhill* was '... *In view ... of the possibilities of accidents ...*'

The operating problems were considerable. The usual weight of a parcel of lead (or later lead ore) sold by the mining company was 100 tons. This needed moving safely from Snailbeach to Pontesbury, ideally in one trip. Except for the smelt works branch, it was downhill all the way to Pontesbury with a ruling gradient of about 1 in 38, making control of braking an important issue. The damp, greasy rail conditions common

Snailbeach locomotive shed c.1925 with *No.2* and *No.4* outside. Photo by F. Moore's Railway Photographs, taken between November 1924 and August 1925. A letter to H. F. Stephens at Tonbridge shows that on 11 November 1924, fitter-driver Bill Crawley needed approval for his price of £1-0-0 to lift the boiler of *Dennis*, and on 28 August 1925 Crawley wrote that he had recently removed the steel boiler tubes from *Dennis*. So this photograph falls between the two dates
The boiler of *Dennis* has been lifted from the frames and rests on timber blocks. The boiler tubes are still in place. Beside the boiler, laying face down on the footplate, is the front tube plate which has been removed from the boiler barrel. The internal, or main, steam pipe can be seen sticking up vertically from it. When in service this pipe, as its name suggests, is inside the boiler, and connects the regulator in the steam dome with the cylinders, via additional plumbing in the smoke box. Looking into the open end of the boiler barrel the longitudinal stays can just be made out; these effectively tie the upper part of the front tube plate to the rear of the boiler. The smoke box stands on the ground beside the engine. Note the white lime scale in the boiler barrel, on the boiler tubes, and on the tube plate and the steam pipe. The site seems quite tidy and, in particular, *No.2* looks clean and well cared for. The front footplate, chimney and smoke box appear to shine as if wiped over with an oily rag. COLLECTION P. INGHAM

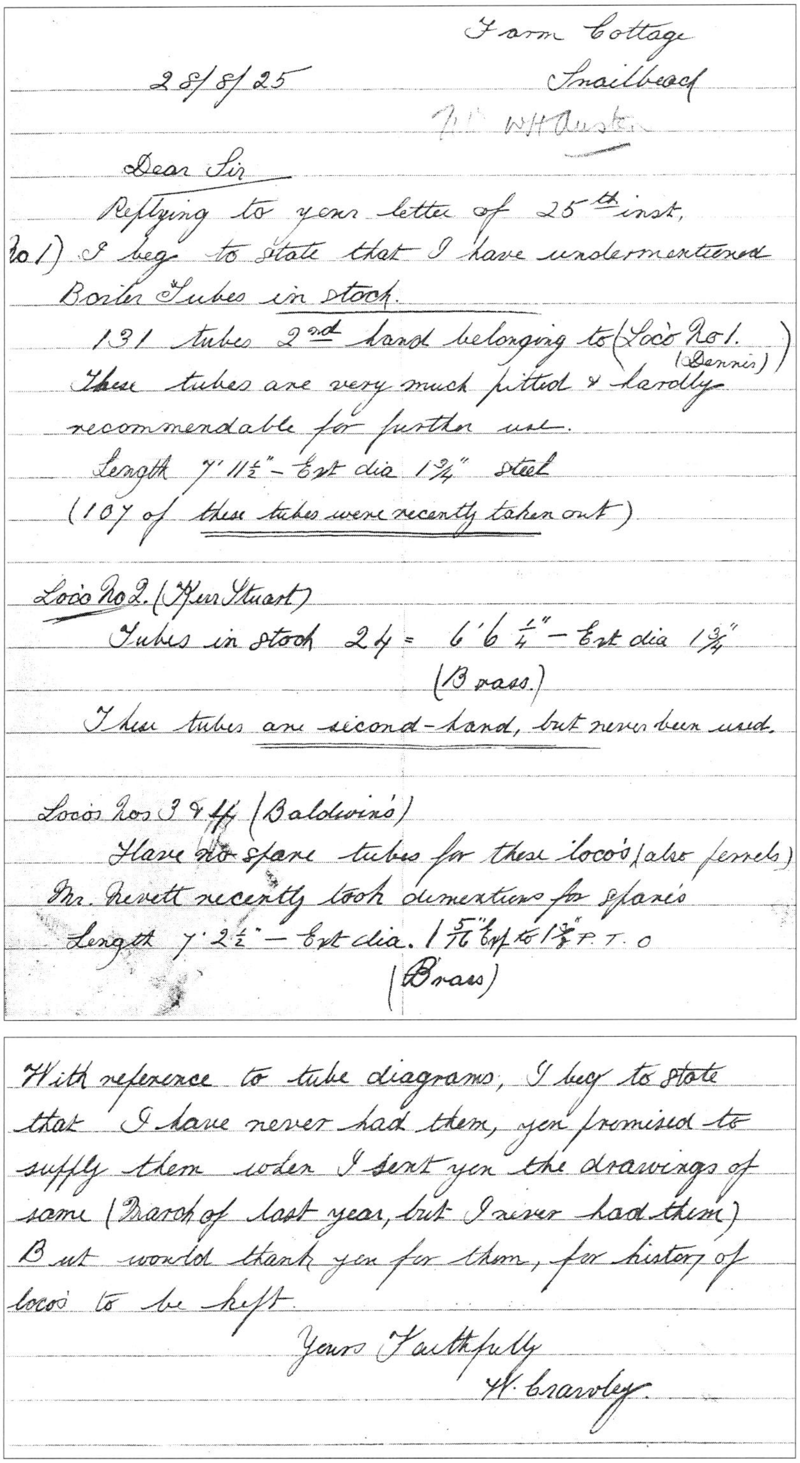
Farm Cottage
28/8/25 Snailbeach
Mr W H Austen

Dear Sir
Replying to your letter of 25th inst,
No 1) I beg to state that I have undermentioned
Boiler Tubes in stock.
131 tubes 2nd hand belonging to (Loco No 1. (Dennis))
These tubes are very much pitted & hardly
recommendable for further use.
Length 7' 11½" – Ext dia 1¾" steel
(107 of these tubes were recently taken out)

Loco No 2. (Kerr Stuart)
Tubes in stock 24 = 6' 6¼" – Ext dia 1¾"
(Brass)
These tubes are second-hand, but never been used.

Locos Nos 3 & 4 (Baldwin's)
Have no spare tubes for these loco's (also ferrels)
Mr. Nevett recently took dimentions for spares
Length 7' 2½" – Ext dia. 1 5/16" Exp to 1 3/8" P.T.O
(Brass)

With reference to tube diagrams, I beg to state
that I have never had them, you promised to
supply them when I sent you the drawings of
same (March of last year, but I never had them)
But would thank you for them, for history of
loco's to be kept.
Yours Faithfully
W. Crawley.

Fitter-driver Bill Crawley's letter to W. H. Austen dated 28 August 1925, concerning boiler tubes and record keeping. COLONEL STEPHENS RAILWAY MUSEUM ARCHIVE

in this Welsh border location would have added to this. *Belmont* weighed only around ten tons, or 13 tons in working order, with only two pairs of driving wheels for traction and braking. Headed by such a small, light locomotive, a train weighing 100 tons or more could have overcome the braking power easily, and perhaps have caused a run-away resulting in a collision with the stop blocks at Pontesbury. This might have been the kind of accident the directors had in mind. It is equally possible that they were considering other 'accidents', such as serious and unforeseen mechanical failures putting *Belmont* out of service, leaving the railway unable to operate as they would have wished. It would not have been good practice to carry on without a spare engine.

Perhaps it was hoped that buying a larger, more powerful locomotive would improve the braking force as well as the haulage capacity. But despite any supposed deficiency of power or braking, the locomotive crew certainly coped with *Belmont* in those early years, because the railway worked and was profitable. It is not known exactly what they did, but there are well known railway practices which they might have used in the down direction. These include sanding the rails, and pinning the wagon brakes down so that the brake shoes drag. The latter was commonly used on down grades on any railway where continuous brakes were not in use. A small amount of power would be applied by the engine to keep the train moving.

Fernhill was better designed both for haulage and to help control the heavy downhill loads. *Belmont*, working from the known locomotive dimensions, and assuming a Mean Effective Pressure (MEP) of 110 lb/in^2, had a Tractive Effort (TE) of about 5,240 lb. *Fernhill* had a TE of about 8,800 lb., and so was about 68 percent more powerful than *Belmont*.[11] This, combined with having about double the adhesive weight, bearing on six driving wheels rather than four, gave it far more effective haulage and braking. Barclays & Co. was fitting steam brakes to its locomotives at the time it was built (as *Dennis* was later), but it is not known if *Fernhill* had them. Essentially, *Fernhill* was a design of locomotive primarily intended for standard gauge, but adapted in this case to a gauge of 2ft 4in, which was an easy way to meet this need. If *Belmont* had been doing the job quite well, *Dennis* would have looked for a similar small narrow gauge locomotive again, at a much lower price. *Fernhill* can be judged a successful design because it was in service for at least 20 years. *Dennis* was in all critical respects equal to *Fernhill*, showing that the high specification was proven and necessary. At the same MEP of 110 lb/in^2, it delivered a TE of 8,575 lb. The heavy 0-6-0 tank locomotive was the preferred type during the lead mining era. The existence of *Belmont* in the stud was because of force of circumstance – in 1876 capital was short. The traffic and operating conditions of the line governed the choice of locomotive when more money was available.

1923 – 1947

The new owner of the railway needed replacement motive power straightaway, and it was fortunate that there were still many war surplus narrow gauge locomotives available. The majority were built specifically for military service, but some had been requisitioned from civilian ownership at the start of hostilities. The SDR had examples of both. Meanwhile, the old locomotive *Dennis* was thought to be worth saving, and the extra locomotives were numbered 2, 3 and 4 accordingly.

Spares were ordered early in 1923 in anticipation of the repair of *Dennis*, but it took until late 1924 for the order to be given for serious work to begin. This was described in a letter of November 1924 from a consulting engineer Nevitt of Llandudno to H. F. Stephens. Fitter-driver Bill Crawley was paid a piece rate for drilling out the rivets to be removed from the copper firebox stays, foundation ring, fire hole door and smoke box tube plate. He was also paid £1 0s 0d to lift the boiler out of the locomotive main frame, which was a relatively large sum on top of his wages. This probably included an amount for hiring men to help with the levering up and packing with timber, and the other heavy work like lifting the cab off. Various photographs show the dismantled pieces laying beside the locomotive shed at Snailbeach. Drawings and sketches of the firebox originating from the Tonbridge office, and from the SDR fitter-drivers Crawley and Gatford, show that the firebox was suffering from general wear and tear, and that half side copper patching had been added years earlier. If *Dennis* was to be returned to service safely this time, a new fire box would have been needed.

As can be seen from photographs, some work was done after Crawley's initial dismantling of *Dennis*. The work described in Nevitt's letter to H. F. Stephens continued, and the boiler tubes were removed by August 1925. Following this, the rear of the boiler and the firebox were also removed, pending repair or replacement, and the lime scale inside the boiler barrel was chipped away, as can be seen in a photograph of 1932. Thinning and perforation of the smoke box is also evident, showing that it also needed replacement.

Intermittent correspondence between H. F. Stephens and Snailbeach about repairing *Dennis* continued until as late as 1929, but the work was never completed. The SDR now had three adequate locomotives in good running order which were coping with the increasing traffic, and there was no real justification in having a new fire box and smoke box made, simply to return another locomotive to service. *Dennis* was beyond economical repair.

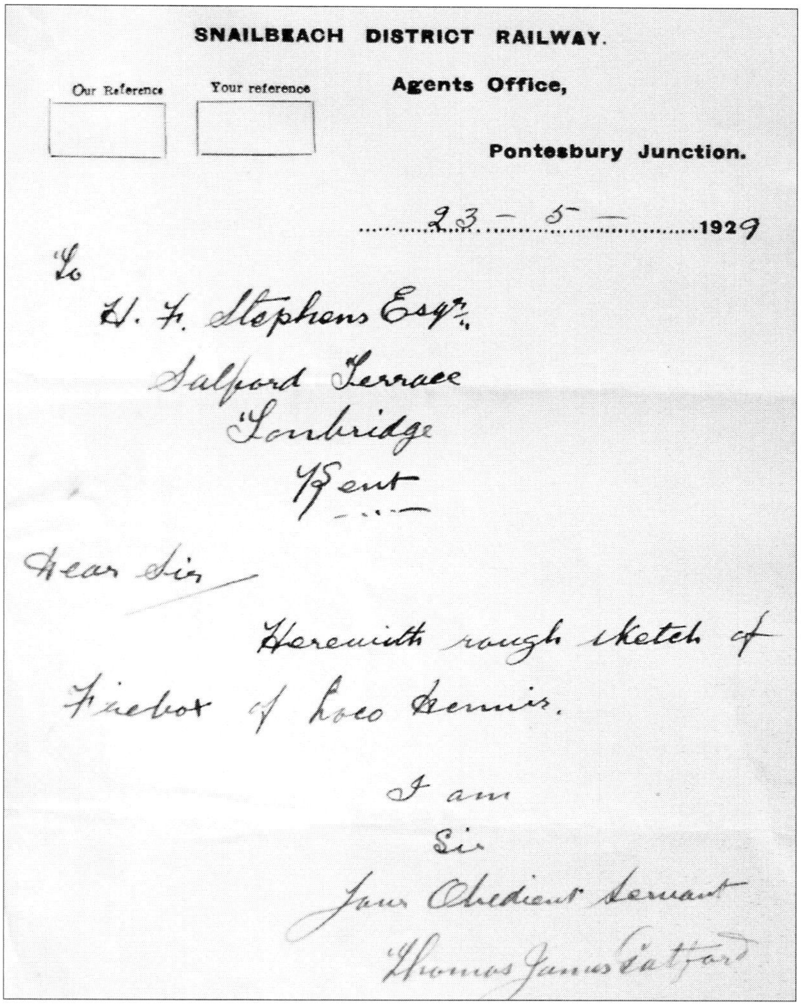

SNAILBEACH DISTRICT RAILWAY.

Our Reference | Your reference

Agents Office,

Pontesbury Junction.

23 – 5 – 1929

To

H. F. Stephens Esqr.

Salford Terrace

Tonbridge

Kent

Dear Sir

Herewith rough sketch of Firebox of Loco Dennis.

I am

Sir

Your Obedient Servant

Thomas James Gatford

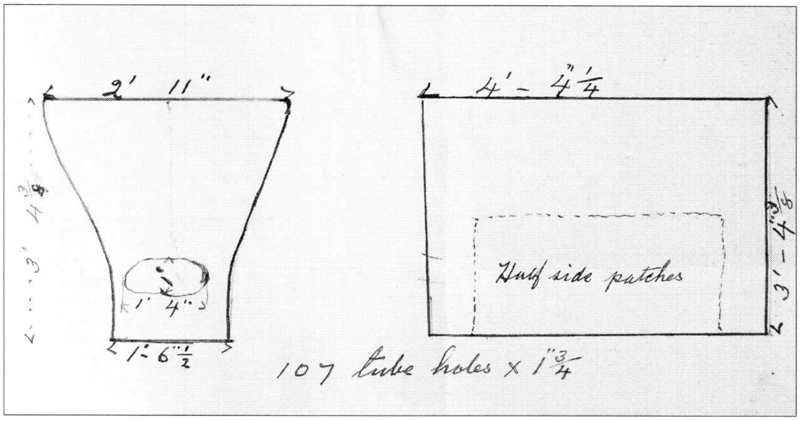

Memo by Tom Gatford regarding his sketch of the firebox of *Dennis*.
And sketch of the firebox of Dennis by Tom Gatford, showing the patches applied at the last major overhaul.
COLONEL STEPHENS RAILWAY MUSEUM ARCHIVE

Above: Rear of cab of *Dennis*, behind Snailbeach locomotive shed, showing cutout for turning the brake standard.
H. F. WHEELLER

Right: The remains of *Dennis* in 1936. Later that year a scrap metal merchant disposed of them.
J. E. SIMPSON
COURTESY P. INGHAM

Stephens fell ill and died in 1931, and with his passing any interest in returning *Dennis* to service was also gone. So the Bagnall stayed in pieces beside the shed at Snailbeach, although the Board of Trade annual returns from the company showed four locomotives until 1935. *Dennis* was officially withdrawn in 1936 and finally consigned to the scrap metal merchant in 1937. [12]

Livery for the working locomotives was black with red buffer beams. Nobel's Stack Black and red lead paints were bought regularly. By the 1930s, the numbers in unknown colour, but

Close-up of *No.2* in 1932. There appears to be little change to the locomotive since the F. Moore photograph c.1925. The whistle projects from the cab roof and has an operating lever on top. The whistle was sounded by pulling a chain or cord, which passed from the lever through the cab roof. The right hand front corner of the cab roof appears always to have been bent up. Either side of the smoke box are two Roscoe displacement lubricators, which were fitted from new. The right hand one can be seen clearly. The two lubricators mounted on top of the steam chests are not original. They were probably robbed from the disused *Dennis*, since a similar lubricator can be seen in the Bagnall works photograph
The smoke box carries a lamp bracket. The door has become a little weak, as can be seen from the smoke staining on the smoke box face. Overtightening in an attempt to make the door air-tight has led to the centre being dished in. This can be seen behind the vertical smoke box door handle and the top of the lower hinge. In an attempt to rectify this, and strengthen the door, a horizontal plate has been fitted, which is held in place by six bolts. This plate can also be seen in F. Moore's photograph. The central buffer and coupling loop survive from Henry Lovatt's Leek & Manifold Valley Railway sub-contract, for which the locomotive was originally built. Note the home-made corrugated iron 'door' fitted inside the cab.
A. W. CROUGHTON

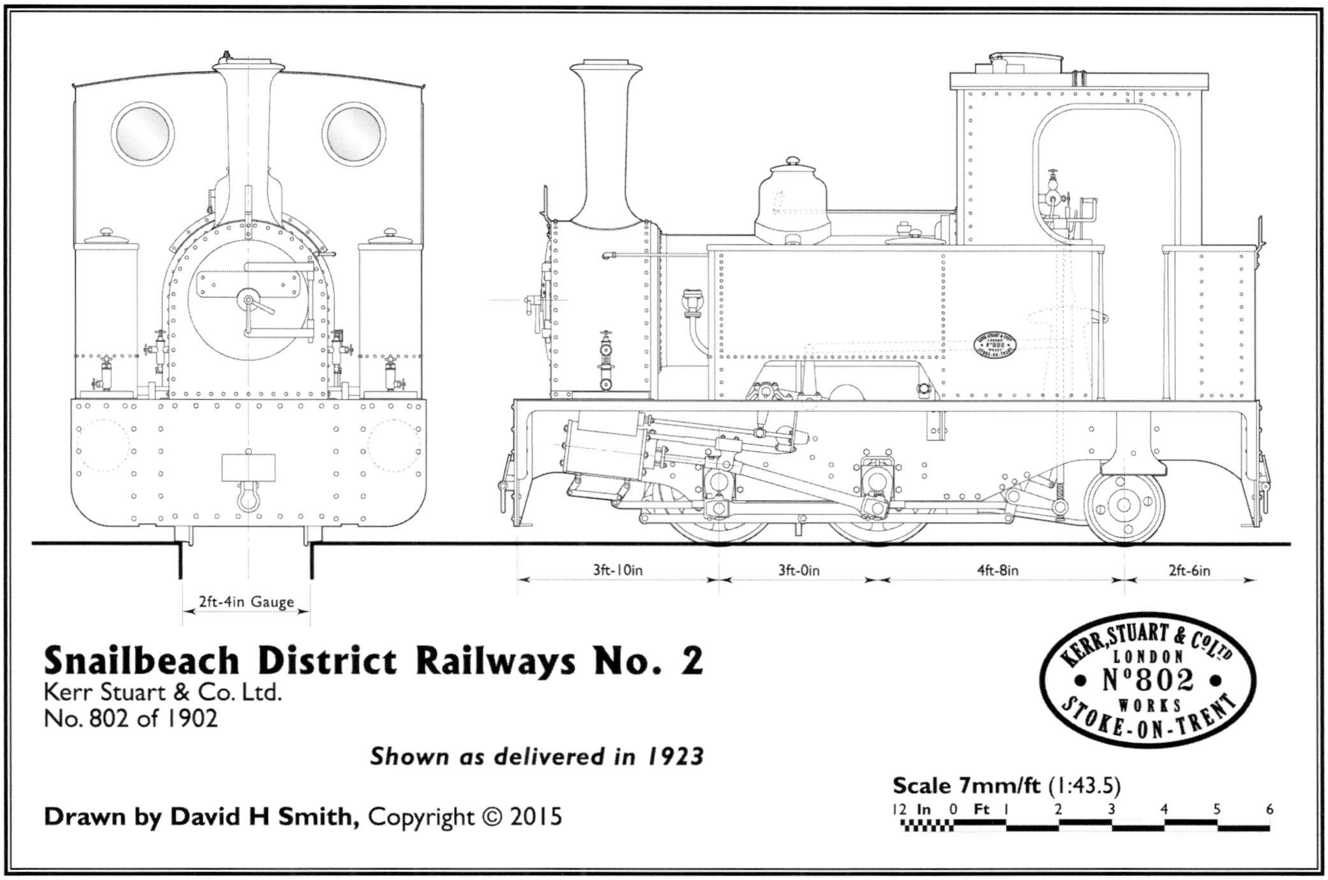

probably white, were added roughly to the buffer beams, as can be seen by the J. F. Bruton photograph of *No.4*. In the 1940s the numbering was quite clearly in white – white lead.

The Kerr, Stuart – *No.2*

This locomotive was nicknamed the *Puff and Dart*, because of its driving characteristics.[13] It was built in 1902 by Kerr, Stuart of Stoke-on-Trent, works number 802, an example of their standard 'Skylark' class. When orders were sparse, locomotives of this type were built to an advanced state and held in stock. Following an order from contractors Henry Lovatt & Co., this example was completed to 2ft 6in gauge. Lovatt was given a sub-contract to build the Leek & Manifold Valley Light Railway in Staffordshire, and the locomotive was delivered there in October 1902. The Kerr, Stuart locomotive register states that standard livery for the class was applied.[14] Other than the class name 'Skylark' being painted on the side tanks, this original livery is unknown.[15]

It is thought that the locomotive then worked on another Lovatt contract, to build No.9 dock at Salford on the Manchester Ship Canal. There was an auction of the contractor's plant for this job in August 1907 which appears to have included Kerr, Stuart 802. Following this, it was engaged on the construction of Horsea torpedo range. In 1914 it was in storage at the Admiralty depot on Hoo Ness Island, Kent, then being sold to J. Wardell & Co. Ltd.[16]

The Ministry of Munitions (MoM) requisitioned the locomotive in about 1916 for Admiralty use. It was used in the construction of the Ridham dock of Edward Lloyd Ltd on the Swale, the tidal waterway separating the Isle of Sheppey from mainland Kent. Ridham was a strategically important General Salvage Depot until it was decommissioned around 1919. The Kerr, Stuart locomotive then became surplus to requirements, and was put in storage ready for re-sale.

The MoM Disposal Board began the sale of war surplus equipment in 1919, and from then until 1923 a catalogue called *Surplus* was published at frequent intervals in a bid to sell all the remaining surplus material of all descriptions. In 1922 George Cohen Sons & Co. Ltd made an agreement to purchase all the

General view at Snailbeach locomotive shed, 1932. Some work has been done to *Dennis* since 1925. You can see through the boiler barrel, so the rear of the boiler, the firebox and all other internal parts have been removed. The longitudinal stays have been laid on the footplate – at least one has rolled to the ground. The boiler barrel has been totally cleared of lime scale. But by now H. F. Stephens was dead, and any prospect of the locomotive being rebuilt had passed. A. W. CROUGHTON

remaining material. Then in 1923 a combination of Cohen's and Sir W. G. Armstrong Whitworth & Co. Ltd took over the activities of the Disposal Board.[17] *Surplus* carried advertisements for Kerr, Stuart 802 in 1920 and 1921 – on the first occasion it was lying at Ridham.[18] It then passed to the dealer E. C. Cornforth of Stoke-on-Trent, who sold it to the Snailbeach District Railways on 11 December 1922. It was delivered to the line on 10 March 1923.

THE BALDWINS – *No.3* AND *No.4*

These two locomotives were built for the MoM by the Baldwin Locomotive Works of Philadelphia, U S A. They were part of a large order for Baldwin class 10-13-D locomotives for use on War Department Light Railways. These were supply lines behind the Western Front in the Great War, chiefly of 60cm gauge. Late in the war, W. G. Bagnall Ltd of Stafford supplied spares for these locomotives. Bagnall also

No.2 in 1936. Note the patching of the smoke box, and the roof top whistle has gone. H. F. G. DALSTON

No.2 in the early 1940s has the new smoke box fitted.

No.2 on 11 June 1943.
L. W. PERKINS

No.2 on 11 June 1943, with its new smokebox and replacement smokebox door. The cylinders no longer have drain cock linkage to the driver's cab. Once the condensation had been cleared from the cylinders, a spanner would be used to close the cocks.
Seen in the background are a number of mine buildings of considerable age. Dominating the centre is the old mine office building, which has parts of different ages and architectural styles at several levels. To its right, and partly obscured by the smoke from *No.2*, is the head frame (for a single rope winding system installed by the Halvans Co.) over George's Shaft. Behind the head frame is the miners' dry. The roof of the dry is shorn of most its slates but still has the louvred ventilator. The imposing skyline of Resting Hill caps the picture. L. W. PERKINS

repaired and rebuilt locomotives sent across the Channel from the Front, because of the evacuation of La Berguette Central Workshops near Ypres. After shipping, locomotives were despatched from MoM dumps at Purfleet to Stafford, and returned to Purfleet on completion, in anticipation of returning them to war service. Several of the locomotives were fitted with a simple cab extension and vacuum brakes while under repair at Stafford. This left a convenient supply of refurbished locomotives suitable for sale to many narrow gauge railways after the war. A number of British railways bought these war surplus machines.

There appears to have been some kind of trading agreement between the Cohen Armstrong consortium and smaller dealers. H. F. Stephens bought the two rebuilt Baldwins, works numbers 44383 and 44522, from Learoyd & Son of Clapton, East London, in January 1923, and they were delivered to the Snailbeach line in April 1923. They became known as *No.3* and *No.4* respectively.

Having been rebuilt by Bagnall, the Baldwins

Some basic facts about the SDR examples, up to completion of the Bagnall rebuilds:

WD Order No.	No. of locos ordered	Date of Order	Date built	Baldwin Locomotive Works No.	Baldwin Serial No.	WDLR No.	Bagnall Order No.	Arrived at Works	Work completed	Cost	S.D.R. No.
DRT 795	45	Mar 1916	Oct 1916	44383	49	538	6883	12 Dec 1918	16 Apr 1919	£175-14-3	3
DRT 887	350	Sept 1916	Dec 1916	44522	78	722	6540	2 Oct 1918	20 Nov 1919	£210-8-8	4

No.3 photographed at Callow Hill on 19 December 1941. It probably has a weak or broken spring, hence the heeling to one side.
J. G. VINCENT

No.3 photographed at Callow Hill on 19 December 1941. Two re-railing jacks are carried. J. G. VINCENT

were generally ready for service but needed re-gauging. To do this, it is likely that each locomotive was delivered by crane onto timber cradles at Pontesbury, allowing the wheels to be dropped out and sent away by rail to a workshop having the equipment needed and railway locomotive experience. The driving wheel tyres were then pushed outwards on the wheel centres using a powerful industrial press. The four solid carrying wheels were pulled off their axles and turned around, to have their inner faces (with a deeper central boss) facing outwards. Therefore the tyres were also reversed on their wheels before remounting them (in this far from obvious way) to a wider gauge. Close inspection of a restored example of this locomotive type which now operates on the Leighton Buzzard Narrow Gauge Railway in Bedfordshire shows that all ten tyres were an interference fit on the wheel centres. No bolts or retaining rings of any kind were used in the original build, allowing this fairly simple conversion work on the SDR pair.[19] On return to Pontesbury, the wheels were refitted. The SDR fitter Mr J. Webb then re-mounted the brake shoes to line up with the wider-spaced wheel treads, and adjusted the axle bearing assemblies ('wedges and liners').[20] The locomotives were then lifted onto SDR track.

This was a relatively cheap solution but was a compromise, resulting in a locomotive which was not quite the correct gauge for the railway. The movement available for re-gauging was limited by two things: firstly the need to maintain clearance between the driving wheel tyres and the inside faces of the coupling rods; and secondly the safe distance the tyres could be shifted on the wheel centres. The work might even have involved machining the backs of the tyres, to make sure there was clearance between the flanges and the check and wing rails at points. Many photographs show that the driving wheels did not overlap well onto the rails as they should in normal circumstances, indicating that the wheel treads were narrow to gauge. This

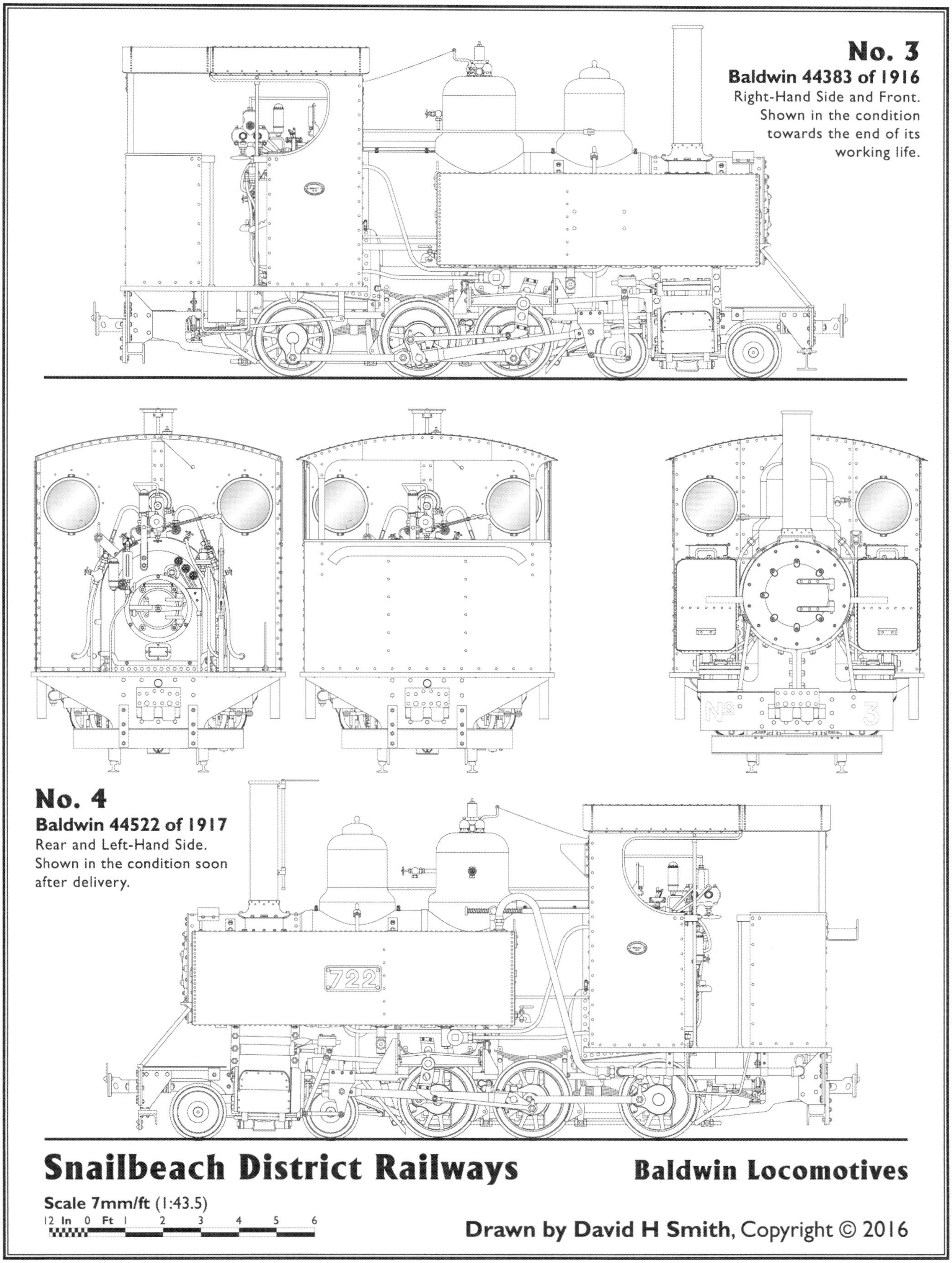
No. 3
Baldwin 44383 of 1916
Right-Hand Side and Front.
Shown in the condition
towards the end of its
working life.
No 3
No. 4
Baldwin 44522 of 1917
Rear and Left-Hand Side.
Shown in the condition soon
after delivery.
722
Snailbeach District Railways
Baldwin Locomotives
Scale 7mm/ft (1:43.5)
12 In 0 Ft 1 2 3 4 5 6
Drawn by David H Smith, Copyright © 2016

No.4 at Snailbeach c.1925. The R.O.D. operating number plate 722 and Bagnall rebuild plate (on the side of the cab) are both carried. The chimney damper flap is fitted, and the steam valve, projecting from the front of the dome, has its take-off facing downwards. This device was not fitted to *No.3*. Note the bogie wheels – on the leading pair the axle appears flush with the wheel centre, whilst the axle on the trailing set appears to be further in. The pairs of wheels might easily have been swapped during earlier maintenance work. Seen on top of the firebox in silhouette is the Detroit sight feed lubricator. The pipe that rises from the rear and top of the tank, before dropping down below the footplate, is the delivery pipe for the water lifter. This device worked in the same way as an injector and was used to lift water from a stream or static tank into the locomotive's tanks, *via* a 2½ inch bore wire-reinforced hose that fitted onto a nozzle located above the rear coupler. Without reinforcement the hose would collapse from the suction when in use. The tanks held 396 gallons and the water lifter could fill these in about 8¼ minutes. When not in use in war service, this hose would be hung on the rack attached to the bunker. At Snailbeach, only *No.4* had the water lifter fitted, and had the hose rack and shields over the cab windows, making identification easy. COLLECTION P. INGHAM

No.4 outside the locomotive shed. When working on it, the balks of timber were being used to support the side rods.
COLLECTION F. JONES, INDUSTRIAL LOCOMOTIVE SOCIETY, COURTESY A. C. BAKER

would have been acceptable for low speed operation. W. G. Bagnall, when doing wheel repairs to these locomotives later, recorded that the tyres were set at a special gauge of 2ft 3¾in, although at what point on the tyres this measurement was made is unknown. In 1946 fitter-driver George Edwards reported his dislike of the Baldwins because having 'narrower wheel treads' they tended to drop between the rails – this problem being a result of the re-gauging compromise.[21]

A feature of the Baldwin design not mentioned in earlier descriptions is that the four-wheeled carrying bogie was mounted on a swing link pivot. This bogie mounting is thought to be the cause of rough riding often complained about with these locomotives. It would occur particularly on rough track. The track of the Leighton Buzzard Narrow Gauge Railway is kept in good order, and no complaints of rough riding there have been made about the restored locomotive mentioned above.[22]

The Stephens policy was to use the three locomotives in rotation. In practice they were used in shifts of two to three weeks. In June to November 1943 this changed to one week shifts.[23]

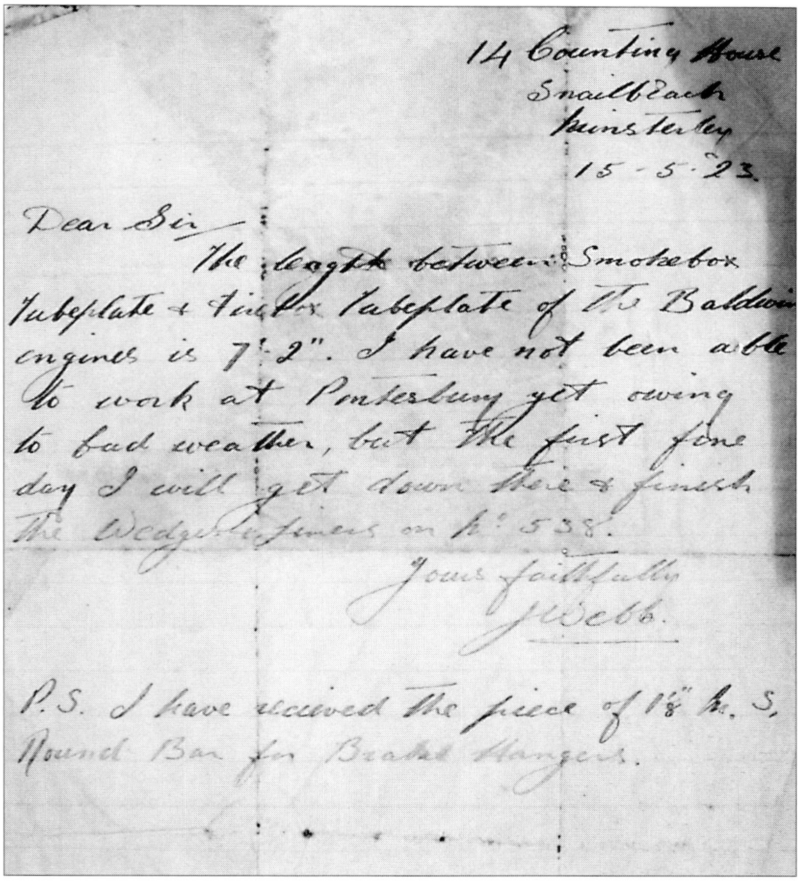

14 Counting House
Snailbeach
Minsterley
15-5-23.

Dear Sir,
The length between Smokebox Tubeplate & Firebox Tubeplate of the Baldwin engines is 7'-2". I have not been able to work at Pontesbury yet owing to bad weather, but the first fine day I will get down there & finish the Wedge liners on No 538.

Yours faithfully
J Webb.

P.S. I have received the piece of 1⅛" M.S. Round Bar for Brake Hangers.

Fitter J. Webb's letter about re-gauging work at Pontesbury.
COLONEL STEPHENS RAILWAY MUSEUM ARCHIVE

Principal locomotive dimensions[24]

Locomotive name	Wheel arrangement & tank type	Cylinder size (in x in)	Leading wheel diameter (ft - in)	Driving wheel diameter (ft - in)	Trailing wheel diameter (ft - in)	Total wheelbase (ft - in)	Boiler working pressure (lb/in²)	Adhesive weight (ton - cwt)	Weight in working order (ton - cwt)
Belmont	0-4-2ST	10 x 15	-	2 - 7½	Unknown	Unknown	140 (assumed)	10 - 0 approx	13 - 0 approx
Fernhill	0-6-0ST	12 x 20	-	3 - 0	-	Unknown	140 (assumed)	15 - 7 approx	19 - 0 approx
Dennis	0-6-0T	12 x 18	-	2 - 9¼	-	8 - 6	150	16 - 10	20 - 0
No.2	0-4-2T	7½ x 12	-	2 - 3	1- 4½	7 - 8	160	8 - 2	9 - 15
No.3 and No.4	4-6-0PT	9 x 12	1 - 4	1 - 11½	-	12 - 2	178	10 - 15	13 - 15

General locomotive maintenance

Until 1883 a full time fitter was employed, who was paid 10s a day. Such a large payment was unlikely to have been a wage, and was probably in the nature of a maintenance contract, where the fitter was expected to hire all the necessary labour and order the spare parts needed to keep the locomotives running. (This was essentially how stationary engine 'engineers' worked on a mine in the 19th century.) This would have been especially important when *Belmont* was the sole motive power. After 1883, specialist contractors were used for major work. These included Cudworth & Johnson of Wrexham. From the 1890s, local mechanics and fitters were increasingly used for general maintenance, and paid at the going rates.[25] Later, some work was done by E. Davies of Shrewsbury: this business was referred to as E. Davies' Atlas Foundry in 1911 in SDRCo correspondence. Davies was listed in a 1917 directory as an *'Engineer, machinist & iron founder; manfr of steam, gas and oil engines etc'*, and also had experience with road and agricultural engines and machinery. The

Snailbeach locomotive shed on a wash day, c.1936. A Bagnall wagon stands on the main siding, in need of repair on its far side. All the weight of the top-hinged side door was taken by a few bolts holding the upper plank in place. These mounting points eventually failed, and the plank is seen propped up in the wagon. *No.2* has its old smoke box, and the remains of locomotive *Dennis* have yet to be sold to the scrap metal merchant. A stockaded stores area stands to the left, to which there are doors on the left hand side. In the background, the tumbledown wall marks the boundary of the old carpenter's yard, where a resident has hung out her laundry. H. F. G. DALSTON

business was incorporated as The Atlas Foundry Ltd in 1944 and refurbished the locomotive *Dolgoch* for the Talyllyn Railway in 1945. Davies had two sites at Frankwell in Shrewsbury.[26] One of these, close to Welsh Bridge, is now occupied by a theatre. The other was the Atlas Foundry itself, which was a few yards further along the River Severn on Frankwell Quay. It was cleared for redevelopment by about 2000, and the new Shrewsbury Guildhall has been built there.[27] The Atlas Foundry's men had the necessary skills and it was very well equipped for road and railway locomotive work, including everything needed for the re-gauging work done on the SDR Baldwins (although it cannot be confirmed that they did the latter work). Machinery seen in the late 1960s included a faceplate lathe having a faceplate at least eight feet in diameter, the lower part of which turned in a pit in the floor. It was used for resurfacing the bearings of road locomotive rear wheels, which could be secured to the faceplate, and could equally well be used for turning railway wheels.

Before 1922, as has been illustrated by the somewhat neglected maintenance of *Dennis*, a combination of poor facilities and perhaps the limited knowledge of the engine men resulted in expensive repair problems. For instance, regular cleaning out of damp ash from the smoke box of *Dennis* would have reduced the potential cost of repair, but this was not done. The heavy scaling of the boiler and tubes resulted from irregular or perhaps non-existent washing out. The low pressure of the water supply at the locomotive shed would have discouraged the necessary washing out of the boiler, which needs water under pressure to remove the scale. In time, the boiler efficiency will have been seriously reduced, so affecting the performance of this or any locomotive.

While day-to-day maintenance of *Belmont* and *Fernhill* will have been covered by the fitter's, driver's and fireman's wages, other sources such as pay sheets, receipts and the occasional Minute Book decision outline some of the major work done on them, without telling us whether they were allowed to get into the same state as *Dennis* did later. These sources give the following details:

Belmont [28]

1882 Major overhaul after working five years with no spare locomotive. £250 0s 0d paid to Lennox, Lange & Co. in May '... on a/c of repairs ...' Two pistons sent from Stafford in August.

1889 Major overhaul by George Watkins & Co. with locomotive out of service for best part of three months. Seven years since previous one – a longer interval with two locomotives now in use.

1892 Locomotive was lifted in June (work unspecified) and in October the Brush

Snailbeach locomotive shed with *No.3* in steam. The maintenance pit can just be seen under the locomotive. Propped up against the door is the frame of the four-wheeled front bogie from *No.4*. The door construction including the original wicket gate can be seen. To the right, *No.4* can be seen jacked up while work is done underneath.
H. F. G. DALSTON

Snailbeach locomotive shed circa 1939. *No.4* stands outside and the old smokebox from *No.2* has been left to one side. The water hose lies next to the locomotive. The siding taken up by *Dennis* has gone and so has the point, but the check rail has been left, and this can just be seen on the right. H. F. G. DALSTON

Electric Co. of Loughborough (now the owner of Hughes's designs) was paid £1 4s 0d for spares.

1895 Five men working on the locomotive in August.

1899 Major overhaul at Cudworth & Johnson's works costing £340 1s 2d.

Fernhill [29]

1888 Firebox repair by George Watkins costing £6 0s 0d, seven years after entering service.

1891 Major overhaul. A bill for boiler tubes and ferrules was received from Andrew Barclay, Son & Co. in October. Hired fitter Charles Phasey repaired the brasses (i.e. bearings) while lodged locally for two weeks in November.

1893 26 man-days were spent painting and cleaning in September, with the work continuing into October. Repairs of 'damage' were done in November.

1895 Four men 'repairing loco engine *Fernhill* with tubes' in December.

LOCOMOTIVE NOT SPECIFIED – EITHER *Belmont* OR *Fernhill*

1894 Repairs to locomotive in August and September.

1895 Repairing locomotive, possibly continuing work on *Fernhill*, see above.

1907 Overhaul costing £86 6s 0d by Cudworth & Johnson of the spare locomotive remaining after *Dennis* was bought.

The pay sheets continue in an undisturbed series through the 1890s to 1903, but with the above exceptions they show no detail of the actual work done on the locomotives except hours worked by the fitters. When *Belmont* returned from Cudworth & Johnson in February 1900, repairs to *Fernhill* were continued until December 1901, indicating that both locomotives were in use at least until then. From that date and until the surviving pay sheets end in October 1903, only repairs to *Belmont* were done, suggesting that *Fernhill* had fallen out of use, but this is not proven. The Board of Trade returns show that the company declared two locomotives throughout this period, but this is no indication of whether or not they were in working order. *Dennis* for example was declared as one of four locomotives on the line between 1923 and 1935, although it was never returned to working order.

After 1922, under the ownership of H. F. Stephens and his legatees, as much work as possible was done at Snailbeach by the fitter-drivers. Spares of all descriptions were bought by the Tonbridge office and delivered at the railway, including wheels and motion parts. Suppliers included W. G. Bagnall, E. Davies and the Ashover Light Railway. If required, help could be sent from the Kinnerley locomotive depot of the Shropshire & Montgomeryshire Light Railway, which was also part of the Stephens group. Kinnerley depot had machine tools, and made and modified spare parts for the SDR locomotives. When needed, a boiler test pump was loaned by the Clevedon depot of the WC&PR.[30]

Maintenance facilities at Snailbeach locomotive shed were rudimentary. There is an inspection pit under each of the two shed roads which are now covered over. E. S. Tonks notes that the shed was equipped with a lathe (date of origin and power source unknown); a smith's furnace (or hearth – again date of origin unknown); some primitive lifting equipment; and general tools for wagon repairs. But from 1947, with the leasing of the lower part of the line to the Shropshire County Council, the upper part of the line was no longer operated, steam operation had ceased, and any repairs to wagons were done at the Shropshire County Council's workshops at Meole Brace, Shrewsbury.

As seen in the 1980s, work benches were provided against the east and south walls, below the three windows. The base of the hearth was also against the east wall (in the north east corner). A flue was provided for the hearth, the remains of which poked out of the neighbouring ventilation slot under the gable for many years. Above the hearth the wall was seen to be heavily smoke-blackened, but this has now been obliterated by the white paint of the renovation (see Chapter Eleven). A corrugated sheet enclosure which cannot be dated was built outside against the north wall, which was probably for railway stores but this is unconfirmed. An earth closet was added, announced by a memo from Austen in May 1933.

24 May 1952. View inside locomotive shed. Note the inspection pit, window openings, work bench and smoke hood.
F. W. SHUTTLEWORTH

Most of these hopper wagons are ex-Granham's Moor Quarry Co. and photographed c.1936 on the west siding. The near siding leads to the narrow-to-standard gauge transhipment pier. The boarded area to the far right marks the start of the standard gauge siding, which is walled on either side for a short distance before the timber built pier is reached.
The external wagon metalwork is generally made from rolled steel strip and plate. The steel hopper linings have protruding bolts. These attach the encasing boards, corner plates and strapping by using nuts fixed on the outside. Shorter couplings are now used which can hang loose without dragging along the track. Rather than being underslung, they are mounted on the headstock faces, between the twin dumb-buffers, which are formed from single strips of steel. They also have sprung draw gear. The brake hangers are trapezoidal pieces of steel plate. Note that two of the wagons have modified hopper doors, which deposit some of the stone outside the rails. The door hinges can be seen bolted to the sole bars. H. F. G. DALSTON

CHAPTER FOURTEEN

Wagons

Original wagons

Wagons on any railway are almost a primary concern, being just as important as the track, goods facilities and locomotives. They deliver goods at agreed rates, and so are a means of getting income. The first wagons on the Snailbeach railway, delivered in 1877, were built by a little known concern named the Union Railway Carriage and Wagon Co. Ltd (URCWCo), of Chorley, Manchester. The vehicles supplied were 30 four-ton coal wagons for unloading by shovel; twelve hopper wagons for unloading by gravity through a bottom trap door; six flat timber trucks; six low sided goods wagons for loading by hand or crane; and one miners' car to seat twenty. Plenty of spare parts were also supplied, including brake blocks and axle boxes, at an all-inclusive cost of £1,100.[1]

The URCWCo was a short lived enterprise. The company was formed by purchasing the Union Wagon Works in 1874, for which the proprietors John Whittle and William Rushforth were paid in shares. Even before the SDRCo had made its final payment for the vehicles in 1879, the URCWCo was in liquidation, and was wound up on 22 September 1880. No further information or records of the company have yet been found. By 1882 a new company called John Whittle Railway Waggon Builders was trading in Chorley.[2]

The general appearance of all the URCWCo vehicles except the miners' car is known from photographs. The Thomas photograph of c.1885 (see page 47) shows that the hopper wagons did not originally have vertical extension boards. In this same picture, behind the row of hopper wagons on the carpenter's yard siding is another vehicle. It is very indistinct, but as its shape resembles no other wagon, it is probably the miners' car. A flat wagon is also shown in this photograph, standing on the long siding.

A line of hopper wagons can be seen in Albert Hewitt's panoramic view of the mine, as can coal wagons in the coal yard (see pp.64-65). Unlike later hopper wagons, the bodies were completely of timber. In the photograph of *Fernhill*, an early flat wagon is seen carrying a length of tree trunk. A low sided wagon is glimpsed at a distance in another view of the mine dating from about 1900.

The second-hand wagons of 1908

The number of wagons increased by six to 57 in

Coal wagons 41, 46 and 37 on the east siding at Pontesbury. The wheels and brake shoes are just about worn out, and the variations in brake levers and door ironwork show a hard life of make-do and mend. Notice the grassy bank descending from right to left in the foreground. This was the route of the coal siding.
H. F. G. DALSTON

Pontesbury, with bogie wagon 57 and four coal wagons (one full) on the east siding. Differences between the coal wagons are worth noting. The second one has a single door, while the other three have double doors. No.48 might have a parking brake on the east side, judging by the silhouette, while the far wagon might have none at all. H. F. G. DALSTON

1908.[3] These were four coal wagons and two goods wagons, bought second-hand for less than £25 for the lot, to replace some that were nearly worn out but not broken up.[4] The design or appearance of these is unknown.

Large capacity hopper wagons

These were similar to the URCWCo hopper wagons but were generally larger, with extra planks on the vertical hopper extensions and longer wheelbase. Some had single buffing pads on the headstocks, underslung draw gear, and no wear-resisting steel lining in the hopper body. More modern examples had hopper bodies with steel linings (but trap doors from timber only), double dumb-buffers formed from steel strip, sprung horizontal draw gear and other parts made from steel rather than iron. It is not known who built any of these wagons, but it is thought that the modern type arrived in 1905 to 1906,

This view of Pontesbury c.1925 has much interesting detail. The derrick in the background is handling timber brought down the line for timber merchant Jabez Barker of Shrewsbury. On the north side of the branch line is a ground frame hut, which has replaced the original signal box, and there are no longer any disc (ground) signals. On SDR land are two stacks of standard gauge sleepers (cut into halves) for track repairs

In the foreground is an older, large capacity hopper wagon on the overhead pier. It has one central buffing pad. The out-of-use coupling is stowed on a hook under the headstock. It consists of a hook, twisted link and shackle. In the cutting stands a coal wagon, opposite the standard gauge turntable-tippler (mid-centre). The tippler was shortly to be removed and sold. A Bagnall wagon stands at the end of the east siding in the right background. COLLECTION P. INGHAM

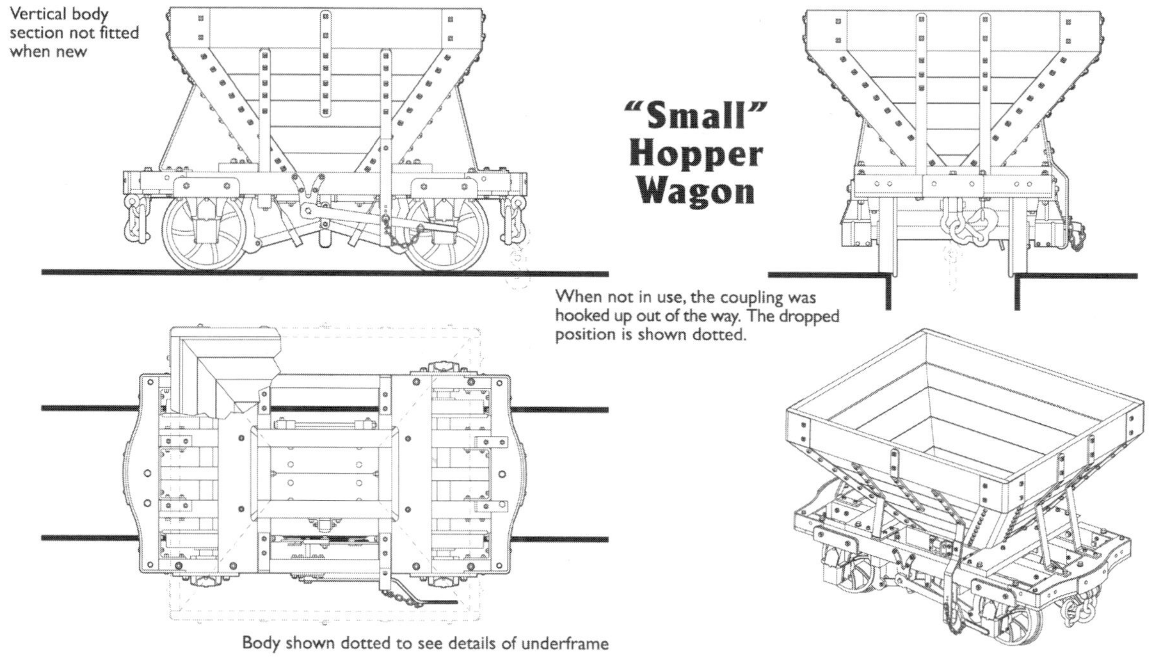

Snailbeach District Railways
Hopper Wagons

These drawings are derived from a variety of sources, including known dimensions and close study of a large number of photographs. They must be regarded as typical of the type, rather than of specific wagons.

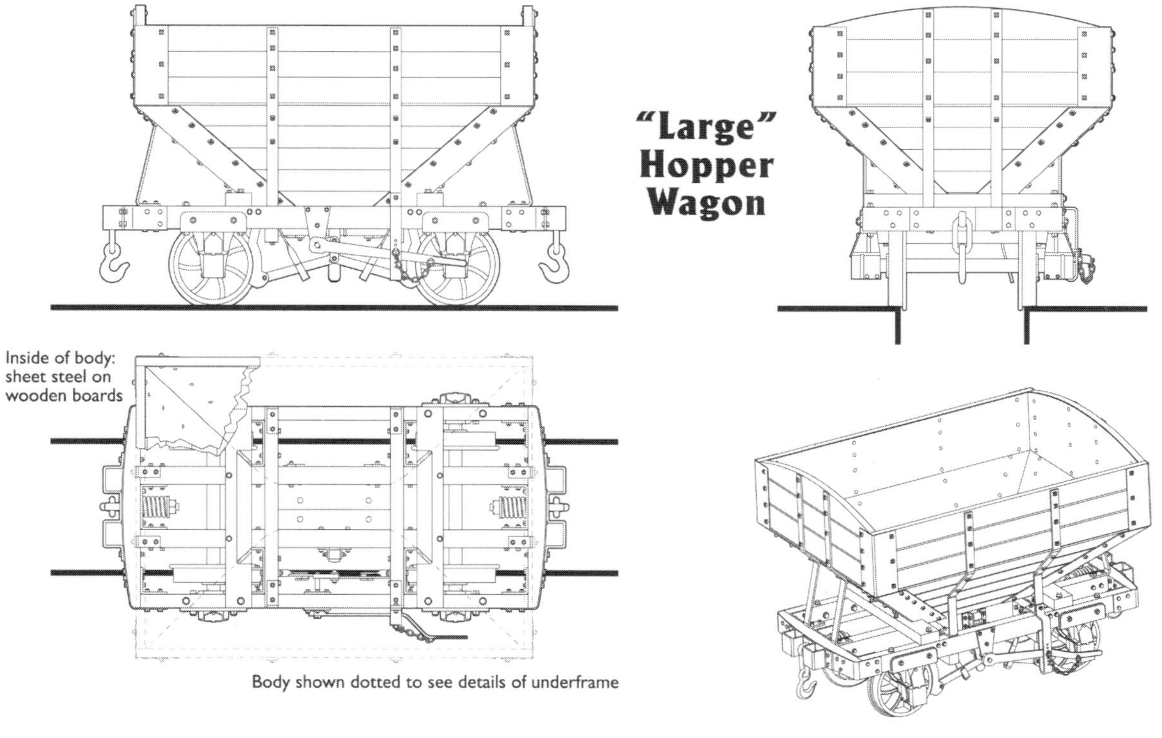

Scale 7mm/ft (1:43.5)

12 In 0 Ft 1 2 3 4 5 6

Small hopper wagons at the end of Pontesbury exchange pier, which was extended for Granham's Moor quarry traffic in 1906. The wagons are much rebuilt and with vertical hopper extensions. The stop block is noteworthy, especially the huge coach bolts fixing it to the last sleeper. I. J. BROWN

when they began privately-owned service with the Granham's Moor Quarry Co. Ltd (GMQCo).[5]

A photograph of the Granham's Moor quarry tramway (about 1906, while under construction) shows that hopper wagons being used were of a modern type, with double dumb-buffers and sprung horizontal draw gear being visible. It is highly unlikely that the GMQCo would be using SDR wagons on construction work. The SDRCo took a mortgage to invest in provisions for the GMQCo contract, which did not include new wagons. So it is likely that the GMQCo's private owner wagons were of the modern type and were bought immediately for the building and operation of the tramway; and the modern examples photographed on the SDR from the 1920s were all ex-GMQCo wagons, 18 of which the SDRCo bought in 1935.

Notable features of the original wagons and large capacity hopper wagons

The URCWCo vehicles and the large capacity hopper wagons had everything which the railwaymen needed to get to – parking brake, travelling hooks for the couplers, securing pins for hopper trap doors – on only one side of the wagon, to save the labour of clambering over them and walking up and down the siding. It also made the wagons less expensive by reducing the amount of iron fittings needed. The coal wagons also had unloading doors on one side only. This was made convenient by planning all the coal unloading facilities on the railway to be on the one side of the sidings which served them. All wagon types were placed on the line at Pontesbury with these features facing west.

This Bagnall design, drawing no.337 for a stone wagon with side and end doors, was first built in 1897 and used on the Plynlimon & Hafan Tramway. The SDR had several of identical design, probably bought second hand, one of which later carried a fuel oil tank at Callow Hill quarry. Drawn from the W. G. Bagnall original by E. A. Wade. COURTESY ELAINE WADE

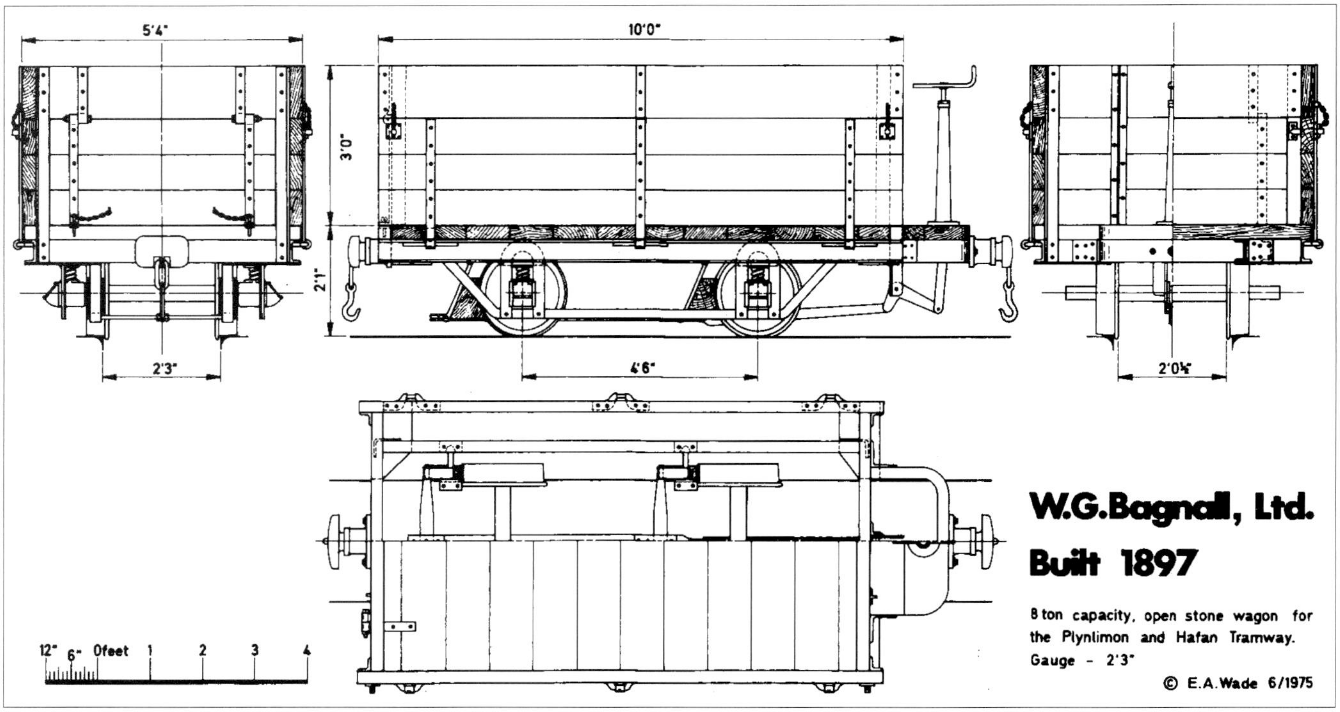

THE SIDE & END DOOR SCREW-BRAKED WAGONS

These were built by W. G. Bagnall Ltd, but there is no evidence of an order from the SDR for them in surviving Bagnall records.[6] Neither is there anything recording their purchase in records held by Shropshire Archives. There were four of this type of wagon supplied to the Plynlimon & Hafan Tramway in Cardiganshire, where they were classed as eight ton stone wagons. A Bagnall drawing of these exists, number W337, carrying the order number 444.[7] How many of these wagons the SDR had is unclear. E. S. Tonks reported two, and perhaps a third was converted to carry the oil tank at Callow Hill quarry, as can be seen in photographs.

THE EXPLOSIVES VAN

A letter of 1913 mentioned a 'powder van' among the rolling stock of 1912. It is not known who the builder was or when it was bought. The van body was metal clad with doors on one side only – interestingly, they were on the east side when seen at Pontesbury. The underframe was similar to those fitted to the later hopper wagons, but with a longer wheelbase.[8] In the four years of detailed returns available between 1913 and 1921, a closed merchandise wagon is listed, but it did not appear in the Board of Trade returns in 1922. After that year it appears almost continuously in the statistics. It was still declared by the company and recorded by visitors in the 1940s.[9]

THE SCREW-BRAKED BOGIE WAGONS

These two wagons were probably bought in about 1923. They were made in Germany to the German specification Modell 18, seeing service on German military supply lines during the Great War.[10] At the Armistice they became 'spoils of war' and were shipped to Britain. Modell 18 wagons are illustrated in fragments of a rolling stock catalogue issued by Honeywill Brothers, a company which traded in ex-War Department railway equipment. These fragmentary pages are part of a collection of relics from the Tonbridge offices of H. F. Stephens, so it seems likely that Stephens bought the two wagons from this supplier. For service on the SDR they were probably re-gauged from 60cm to suit 2ft 4in gauge by pulling the wheels outwards on the axles, with the brake gear modified to suit. Honeywill detailed this type of wagon in a variant having lift-off sides and ends as originally built, and another variant with hinged drop sides, both only in 60cm gauge.[11] The SDR examples were painted red, which suggests a red lead finish.[12] Red lead was bought regularly for the railway stores.[13]

Small hopper wagon No.1 on the overhead pier. COLLECTION D. A. BREWER

THE FOUR-WHEELED DROP-SIDE WAGON

Only one wagon is known of this type – No.45 – and through photographic record only. No.45 does not appear to have any construction or hardware features in common with other supposedly early wagons such as the small hoppers and the coal wagons. For instance, it seems to have had plain axle bearings rather than the usual grease-boxes.

24 May 1952. Underside of small hopper wagon on Pontesbury exchange pier with trap door open. They were originally built by the Union Railway Carriage & Wagon Co. Ltd. Many design and constructional details can be made out in this view. F. W. SHUTTLEWORTH

This end three-quarter view of a flat wagon shows the old-style buffing plate and the unsprung draw gear bolted under the headstock. Notice that everything which the shunter needs to get to – parking brake, travelling hooks for the couplers – is always on the same side of the wagon, to save labour. It also has an original wrought iron brake hanger. C. PEALLING

LATER FLAT WAGONS

Photographs taken in the 1940s and 1950s show flat wagons similar (as far as underframe construction is concerned) to the large capacity hopper wagons. But another type of flat wagon – No.59 – is shown in some photographs. It was very similar to slate slab wagons found on north Wales quarry railways. The maker or source of these wagons is unknown.

This top view shows similarities and differences between an old flat wagon and a modern hopper underframe. The wider platform is provided by adding outer stringers on wider headstocks – other than this the two underframes are essentially the same. Note the reverse-curve headstock profile, similar to other old wagons. On the right, the mounting for the brake lever guide is seen, minus a bolt. C. PEALLING

TANK WAGONS

Fuel oil was needed at Callow Hill quarry for the Ruston Bucyrus engine which powered the crushers and screens.[14] For this, a fuel supplier's storage tank was mounted on flat wagon No.51. This wagon appears to have been a Bagnall open wagon with the superstructure removed, and when used as a tank wagon was stationed on a short length of track at Callow Hill. In the 1940s the tank was marked Royal Daylight, a trade mark of Esso – and also had the number 42 painted on it. In the 1950s an Esso tank was used. There was also a rectangular tank wagon (purpose unknown), the earliest dated record of which is a 1947 photograph. This confirms that the two tank wagons existed at the same time. No tank wagons, identified as such, were ever officially declared or listed as SDRCo vehicles.[15]

FLEET SIZE, WAGON TYPES AND THE BOARD OF TRADE RETURNS

From 1877 to 1912, the Board of Trade returns tell us almost nothing about the company's rolling stock other than an indication of the total number. This is because they have only two general categories for recording numbers of wagons – either *'Wagons of all kinds, used for the Conveyance of Livestock, Minerals, or General Merchandise'* or *'Any other carriages or wagons used on the railway, not included in the preceding columns'*. The only other rolling stock category is

A close-up view of an ex-Granham's Moor Quarry Co. hopper wagon, showing the steel parts of the brake gear and the cast axle boxes. C. R. PEALLING

'Passenger Carriages Only'. More is known from URCWCo's invoice, showing what was delivered in 1877, and the Sale Particulars of 1894 when the Shropshire Mineral Railway Co. was negotiating the purchase of the line. There appears to have been no change in the numbers and types of wagon between 1877 and 1894, other than the odd clerical error in reporting. The balance of numbers between each class or type of wagon reflects the traffic. The need for more coal wagons than any other type is explained by the need for fuel for the lead smelting furnaces, and for the various stationary engines at the mine. Board of Trade figures show that in 1878, the first full year of operation, the SDR carried total mineral traffic of 12,637 tons. This is not broken down further, but Mining Statistics show the lead produce of Snailbeach mine as 1,875 tons. Barytes (barite, or barium sulphate) output was 836 tons, and zinc ore 188 tons. So up to three-quarters of the total was coal, although this ignores any traffic from other mines. The balance changed somewhat after 1892 when smelting stopped.

19 December 1941. Bogie wagon 57 with (partially seen) flat wagon 59 to the left. There is much in this photograph to suggest that the sides of No.57 were originally lifted on and off, like No.56. J. G. VINCENT

Bogie wagon 56 with flat wagon 59 to the right and a coal wagon to the left. Although this is a poor photograph, it is the only one known positively to show No.56. It has sides similar to No.57 but without hinges – the sides and ends were lifted off if required. Flat wagon 59 seems to have been an odd-ball. J. I. C. BOYD, COLLECTION P. INGHAM

In 1894 there is the detailed record given in the Sale Particulars, and in 1913 the Board of Trade introduced an improved classification, but for the intervening years (when changes in traffic occurred), there are no further company records and the old classification continued. So for this period, possible changes to the balance of wagon types to match traffic remain practically unknown. But two changes in the wagon fleet are on record. Replacement wagons were bought in 1908 (see above). Then in 1912, the declared numbers of wagons in use dropped by 13 to 44. This is contradicted by a 1913 report that numbers in 1912 were still 57, including 30 coal wagons, 11 hoppers, 7 timber wagons, 8 goods wagons and a powder van – notwithstanding clerical errors.

In 1913, the Board of Trade introduced five sub-classes of wagon: (1) *Merchandise open and* (2) *closed*; (3) *Mineral*; (4) *Specials flat (rail and timber)*; and (5) *Specials other*. Since 1894, many merchandise wagons had been taken out of service leaving only four, but mineral wagons had increased to 17. The total dropped sharply again to 26. Whichever figures were correct in 1912, the trend in overall numbers was decidedly down, but with mineral wagons increasing. In 1892 smelting stopped, and lead ore was shipped out in loose, crushed form. The demand for other minerals found at Snailbeach – such as barytes – began to increase, although tonnages were low. Then in 1905, the GMQCo began shipping its road stone and some coal for its stationary engines via the SDR. By 1912 several stationary engines at Snailbeach had fallen out of use for various reasons. The details given in the contradictory report of 1913 strongly suggest that the numbers and types of wagon present (in or out of use) were much the same as they had always been, but those actually in service were carrying freight for which they were not originally intended – for instance some wagons declared earlier to be 'merchandise wagons' were now known as 'mineral wagons'. This is known to have happened in the 1930s (see below).

The GMQCo bought wagons for use on its tramway and on the SDR, of both hopper and side tipping types. They are first noted early in the Great War, although as mentioned above they were probably bought at the time of construction of their tramway.[16] Stone traffic from that company increased considerably during the war, but having its own wagons it relied less on the SDR fleet. Only 17 wagons in total were declared by the SDR in 1919. In the eight years to 1919 forty of the SDR fleet had become disused. The

The braked bogie of one of the Modell 18 wagons, being used during the 1950 track lifting. A. BARRELL

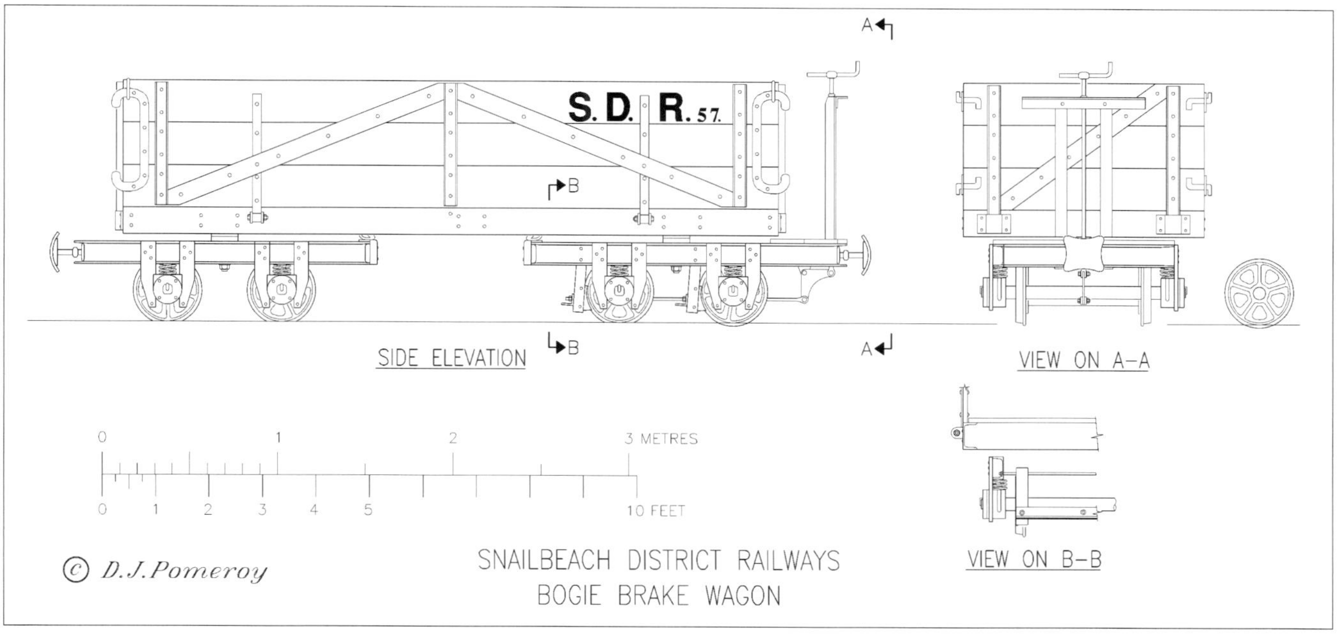

'Modell 18' wagon No.57, drawn by D. J. Pomeroy using known approximate dimensions and the few available photographs. No dimension can be guaranteed as accurate.

SDR needed to hire wagons from the GMQCo as well as repair its own. The increasing scarcity and cost of both labour and materials during the war created extra pressure on the SDRCo to do only the most essential repairs (see Chapter Seven). On average, barely half of the wagons in service received attention in every year after the war, making the repair rate about once every other year per usable wagon. Wagons seem to have stood in various states of disrepair in the sidings, having been abandoned as traffic declined.

The fleet and its condition 1922

In 1921 the usual tonnage capacity figures of 4 tons per mineral wagon, merchandise wagon and flat wagon were entered in the Board of Trade returns. By contrast, in 1922 Dennis & Son saw no reason to maintain the railway and spent almost nothing on wagon repairs. The finish of the Granham's Moor stone trade in 1922 and the general rapid decline in traffic after the Great War resulted in a net loss of £91 for the year, and the sale of the line to H. F. Stephens was expected. Capacities of one ton per mineral and merchandise wagon and a mere half-ton per flat wagon were entered in the Board of Trade returns by Dennis & Son. The reduced tonnage capacity of the wagons indicates their poor condition – or at least the lack of confidence of Henry Dyke Dennis, the line's engineer. He refused to sign a safety certificate for the 1922 Annual Report.

Explosives van 58 and low sided wagon 45. The van was not one of the original fleet. A first glance would suggest that the body was simply dropped onto a hopper wagon chassis, but this was not the case. No.45 was not original either. The non-standard axle-boxes on the latter should be noted. J. I. C. BOYD, COLLECTION P. INGHAM

Rectangular tank wagon and hoppers. The tank wagon must date from the Callow Hill quarry era. It is unusual in having no parking brake, but it has modern dumb buffers and couplers. The small hopper next to it does not have them, because the coupling hook can be seen stowed under the headstock. The underframe of this hopper is breaking up. Again there is a variety of brake hangers on show.
R. K. COPE, COLLECTION P. INGHAM

Dennis & Son, on handing the business over to Stephens late in 1922, provided only twelve serviceable wagons. This was all that was needed to cope with the total 1922 traffic of a little over 3,000 tons – about 60 tons or 20 wagon loads per week. These wagons consisted of four open wagons, four mineral hopper wagons and four flat trucks.

Repairs and renewals 1923-1939

With the Stephens take-over in 1922-1923, the number of serviceable vehicles increased by 19. Merchandise wagons increased to nine and mineral wagons reverted to 17. There were six renewals. Excepting the likely addition of the bogie wagons it is impossible to tell which types they were, and whether or not they were bought. The term 'renewal' in the Board of Trade returns should not be assumed to mean that those wagons were bought new. Heavy repairs for the year totalled 22 and light repairs five. Six wagons were either under or awaiting repairs in December 1923. The company employed a carpenter, who was repairing wagons in May 1923. Hopper wagons were the priority, followed by flat trucks. All existing customers needed hopper wagons for discharging into standard gauge wagons at Pontesbury.

The refurbishments continued in 1924. The number of mineral wagons doubled to 33: a further 16 mineral wagons returned to service by way of heavy or light repairs, but there were no renewals. Stone traffic on the line restarted in 1924 and increased demand. Another mineral wagon was added by renewal in 1927. Old coal (or box) wagons, having side doors, were used for carrying stone for Hayward's Quarries, and reclassifying them as mineral wagons would account for the large increase in this class without buying new wagons. Give or take one, the number of mineral wagons remained static until 1938. Merchandise wagons settled down again to four: six open wagons fell out of use. In 1935, 18 four-ton hopper wagons formerly owned by the GMQCo were bought from the British Quarrying Co. Ltd. This purchase directly follows the agreement under which the Shropshire County Council rented the land beside Pontesbury sidings and installed the tarring plant, which resulted in a renewed demand for hopper wagons. There was no change in the declared rolling stock numbers for 1935 nor in the number of renewals; but new wagon serial numbers appeared in the stone transhipment books. Old box wagons pressed into service as mineral wagons again fell out of use, being replaced by hopper wagons, explaining the change in serial numbers recorded in the books. With only the odd exceptional year the fleet was kept in roughly the same running order to the end of the 1930s.

The 1940s

The fleet in service was almost constant compared to the 1930s. In 1938 for instance, 32 mineral wagons were declared. In the 1940s the mineral wagons were invariably declared at 32, but there were five or six out of use. Enthusiast visitors to the line tended to count every wagon standing in the sidings, out of service or not. On this point, the discrepancies between these numbers and the recorded fleet numbering should be noted (see below). Only three or four open wagons were in service and used for locomotive coal, out of 14 four-wheeled open wagons noted. The ten not in service had finally been abandoned in 1935. The explosives van was in use. The bogie wagons were present but out of use. The tank wagons were not declared as company rolling stock. One of them stood on a short, isolated length of track next to the private siding at Callow Hill. G. Lloyd reported that only 12 wagons were in use at the takeover of the line by the SCC, and that it was still true in 1954.[17]

Repairs and rebuilding

Before 1900, wagon repairs and modifications were let to skilled workers living locally, in the same way that mine maintenance was taken care of. In 1881 for instance, several months of work was done by carpenters in repairing and '*Building New Wagons*'.[18] With regular gravity working, brake blocks often needed replacing. It was common practice to send a good brake block to an iron foundry as a pattern and have spares cast in batches. E. Davies of Shrewsbury was one such supplier. Early in the ownership by the Stephens group a carpenter Jones was employed to put the wagons in order. In the late 1930s a carpenter named Giles was used. There was a carpenter Jones working at Kinnerley on the Shropshire & Montgomeryshire Light Railway. It is highly likely that many of the old original wagons were many times repaired and rebuilt, surviving to see work in the 1930s and 1940s. At the SCC takeover in 1947, repairs were transferred to the Council's works at Meole Brace, Shrewsbury, where wagons were taken to and returned by lorry.

Livery

Nothing is known of the early wagon livery, although lettering was applied in white, in a style much as it was in later years. In the 1930s the wagons were painted stone grey and lettered in white, although occasionally some were not lettered on the east side.[19] However, earlier post-1922 purchase records do not list grey paint but red lead is listed. The bogie wagons were recalled as being red, as noted above. So too was a hopper wagon body which survived upturned in the Pontesbury undergrowth for many years, seen by the author in 1971. Eric Tonks noted that all wagons were grey (as did another author, James Boyd) but that the bogie wagons were originally red. The late Johnny Butler of Snailbeach also said that the bogie wagons were red.

Wagon dimensions

Only the maker's drawing for the W. G. Bagnall wagons is known to survive. Several sources of dimensional information have been found for other wagons, but as none of these can be confirmed against data from the original builders, no dimension can be said to be accurate. Careful examination shows that in some instances dimensions do not correlate between sources.

Dimensions of original wagons

The following dimensions were quoted in the first schedule to the 1894 sale particulars (see Appendix 2 for verbatim descriptions).

Wagon	Description in sale particulars
Miners' car	One seated car (to carry 20 persons) 12ft 0ins x 5ft 0ins
Box (coal) wagons	Thirty-one [*sic*] timber frame Wagons 8ft 8ins x 4ft 8ins x 3ft 0ins internal dimensions
Small hopper wagons	Eleven [*sic*] Hopper Ore Wagons 5ft 6ins x 5ft 2ins x 2ft 8ins bottom discharging aperture 2ft 0ins x 1ft 0ins
Low sided wagons	Six Wagons 8ft 8ins x 4ft 8ins x 1ft 4ins
Flat wagons	Six Timber Wagons 7ft 0ins x 5ft 0ins

Noting the similarity of dimensions of the box and low sided wagons, the dimensions of the latter can probably be taken as being internal. With two inches allowed for timber thickness, both types will have had the same superstructure width of five feet as the miners' car and the flat wagons. The general hopper wagon dimensions are difficult to interpret, but the 5ft 2in dimension might be the inside width of the upper superstructure. It can be compared to the hopper body dimensions given by R. E. Tustin (see below) which are clearly of a small hopper wagon.

Bogie wagon dimensions given in the Honeywill Bros catalogue	
Body internal dimensions	12ft long x 4ft 6in wide x 2ft 3in deep
Distance between bogie centres	9ft 0in

Measurements by R. E. Tustin

The following measurements were recorded by railway modeller Ray Tustin no earlier than the 1940s. They are given in feet and inches. The gen-

Vehicle	Body	Wheelbase	Wheel dia	Length overall	Height	Bogie wheelbase
Flat car 59	6 – 0 x 3 – 5	3 – 4	1 – 5	-	-	-
Bogie 56	12 – 0 x 4 – 9	10 – 9	1 – 0	15 – 0	-	2 – 3
Hopper	7 – 0 x 5 – 6	4 – 6	1 – 6	8 – 6	5 – 6	-
Open 39	10 – 0 x 5 – 6	4 – 6	1 – 6	11 – 4	5 – 0	-
Van 58	9 – 0 x 5 – 0	5 – 9	1 – 6	-	To eaves 7 – 0 Overall 8 – 0	-
General height rail to buffer beam: 2 – 0						

eral buffer beam height is questionable, noting the variety of rolling stock builders and details such as makeshift buffing plates on the bogie wagons.

Measurements by J. I. C. Boyd

As recounted in his book *Saga by Rail*, James Boyd visited the line several times in war time and afterwards, and recorded some wagon dimensions.[20] Where these add to the above information, they are as follows.

Coal wagon 43	Wheelbase 5 – 6
Van 58	Bottom of sole bar to top of roof 6 – 6 Wheel diameter 1 – 7
Open wagon 45	Length 9 – 6 Width 5 – 6 Wheelbase 4 – 6

Fleet numbering

As noted by E. S. Tonks around the 1940s, the wagons were numbered as follows:

Hopper wagons	1-36 (21 supposedly steel bodied)
Coal wagons	37-50 (39 and 42 by W. G. Bagnall Ltd with screw brake and end platform)
Flat wagons	51-55 (51 by W. G. Bagnall Ltd)
Bogie wagons	56 (lift-off sides and ends) and 57 (drop sides and lift-off ends)
Explosives van	58
Flat wagon	59

26 July 1952. A top view of the Royal Daylight diesel oil tank wagon at Callow Hill quarry. F. W. SHUTTLEWORTH

Wagon No.51 later carried the Royal Daylight or Esso oil tank – but note that the Royal Daylight tank itself had the number 42 painted on it in the early 1940s. Nos.1-5, 7, 11, 12 and 16 were small hoppers. Tonks also notes renumbering, giving examples 39, 42 and 47 as being previously numbered 22, 20 and 24. In the Shropshire County Council era of the 1950s different numbers could be found on either side of some wagons, because of cannibalisation during repairs. Some wagons carried no numbers at all then. The lessee honoured the repairing clause of the lease but wagon numbering was of no importance.

Railway wagons, like any kind of freight transport vehicle, from the earliest bullock cart to the most modern articulated diesel lorry, were and are a means to an economic end. This survey of the rolling stock running on the Snailbeach line has covered the various types or classes of wagon, the principal freight they carried, and even changes to the freight they carried and means of loading and unloading. Built-in labour saving devices have also been pointed out. Hire of extra wagons by the SDRCo from the Granham's Moor Quarry Co., when they suited the traffic and when it cost less than repairing out-of-service wagons, was also known, and this is more fully covered in Chapter Seven. All such points fed into the economy of the railway, and were at times critical to the profitability or even survival of the railway company.

CHAPTER FIFTEEN

Managers and men

PRINCIPAL SHAREHOLDERS, DIRECTORS AND MANAGERS

The Lovetts and Joneses – directors, secretaries and major shareholders

The senior shareholders and directors of the Snailbeach Mine Co. Ltd, and the Snailbeach District Railways Co. (SDRCo) in its early years, were the Lovetts and Joneses. All of the mine company's directors were legatees of successive wills, giving them shares in the business. The board confirmed at the first meeting of the Railways Company was essentially the same as that of the Snailbeach Mine Co. Ltd.

The private Snailbeach mine partnership founded in the 1780s originally had ten shareholders.[1] These included three Lovett brothers, Thomas, William and John. Only John Lovett had grandchildren, with the result that the large family interest in Snailbeach mine became concentrated mostly in the hands of his son Joseph Venables Lovett and grandsons Thomas Heaton Lovett and John Henniker Lovett.[2] These three were the core of the partnership in the late 1860s, when it was forced to apply for a private Act of Parliament to reconstitute as a Limited Company. J. V. Lovett died in 1866, and when the Snailbeach Mine Co. Ltd was formed in 1867, Thomas became chairman with John also continuing on the board.[3] Thomas and John then became central to the SDRCo.

John Jones, a landed gentleman of Oswestry, was not an original partner of the old mine partnership, but became the senior partner during the early nineteenth century through bequests and trusteeships. He was associated with several of the partners, both through his business, through marriage and as a personal friend. He was born in about 1761 and died on 8 June 1846.[4] Two generations of John Joneses followed him. His third son, born in Oswestry in 1795, was the most senior John Jones involved with the railway and known as John Jones of Shellbrook (or Shelbrook - his home near Dudleston). He died on 26 May 1874. [5]

The son and heir of John Jones of Shellbrook was born in Oswestry in 1840 and known firstly as John Jones of Higher Grange (Ellesmere) and later of Mossfields (Whitchurch).[6] John Jones of Mossfields maintained the family shareholding in the railway until the 1899 buy-out but had no other involvement. He died in 1923. As well as serving as company secretaries to the SDRCo, the Jones family owned the Jones Mercantile Company, which traded in coal, lime and building materials at Snailbeach. These goods were sold to the mine and railway companies at a discount, and to the public.

An unrelated John Jones, born in 1819, trained in the legal profession and became an Articled Clerk with Longueville's by 1851. By 1861 he was described as an Attorney and Solicitor, and became SDR company secretary in 1875. He died in 1886, and was succeeded as secretary by Edward Williams.

Without the determination and investment of the Lovett and Jones families, and their understanding of what difference Dennis's spending plans would make to the future prospects of Snailbeach mine, there could have been no Snailbeach District Railways Company.

Henry Dennis (engineer; later director and chairman) was a civil and mining engineer who became a successful and prominent industrialist.

Photographic portrait of Henry Dennis. COURTESY DENNIS RUABON & J. MILNER

He was born in 1825 at Bodmin to Matthias and Maria Dennis, and was the fifth of eight children. Matthias Dennis was a successful mercer and clothier who did business all over Cornwall, employing as many as thirty to forty men. Matthias died in 1831 aged only 46, yet Henry's upbringing was provided for, and he attended Bodmin grammar school. He trained under the Bodmin borough surveyor, then joined the engineering staff of the Cornwall Railway. In about 1850 he was appointed by mining engineers John Taylor & Son to construct a tramway from the Oernant slate quarries to the Shropshire Union Canal near Llangollen.[7] After a short spell building mineral railways and lead mining in Spain, he returned to Denbighshire and managed Bryn-yr-owen colliery, near Rhosllannerchrugog, again for John Taylor & Son. In 1857 he formed a partnership of surveyors and mining engineers with Walter Henry Glennie which lasted until 1870. Dennis lived at Hafod-y-bwch from 1857, and in 1859 he married Susan Hicks Stephens of Lostwithiel (Glennie married Susan's sister Lucy a year later).[8] The Dennises moved in 1878 to New Hall, Ruabon, which Dennis bought from Colonel Cornwallis West. They had six children, of which three died in infancy. The survivors were Susan Dewan (born 31 January 1862), Henry Dyke (born 10 November 1863) and Walter Pen (born 26 December 1864).

His obituary in the Proceedings of the Institution of Civil Engineers adds more to our knowledge of Dennis's railway, tramway and mining engineering experience. He is reported to have built mineral railways in southern Spain in 1856 and 1857, which then brought him into contact with the mining branch of civil engineering, within which he was subsequently most associated and grew prosperous. However the dates quoted for his Spanish railway work are perhaps in conflict with the previous details. It might be questioned that he was in Spain in 1857 when he settled in a new home in north Wales in the same year. But without primary source material it is not possible to confirm a chronology for his work in the 1850s and it seems unlikely that such material will now be found. For instance, the known date of his building of the Oernant Tramway is only approximate (based on a date stone), and without the historical records of John Taylor & Son, which were lost in the London Blitz in the Second World War, it might never be known.[9] The Institution of Civil Engineers obituary also reports he was the engineer responsible for the construction of the Wrexham District Tramways: this is confirmed by the Deposited Plans (see Chapter Three).

Dennis built himself an extensive career and range of properties in extractive and manufacturing industries, mainly in Denbighshire, but also in Merioneth and at Snailbeach mine. He was also a managing director of the Glyn Valley Tramway Co. and a director of the Minera Mining Co. At the height of his career the various concerns under his control employed over 10,000 people. He was elected an Associate Member of the Institution of Civil Engineers in 1881, Chairman of the North Wales Coalowners Association, and was elected president of the Mining Association of Great Britain in 1901. He also served as a Justice of the Peace for Ruabon Magistrates' Court. A Conservative in politics, he declined invitations to stand as a candidate for Parliament. When the Denbighshire County Council was formed he was elected one of the first county aldermen 'in recognition of his wonderful business aptitude and as a pioneer and captain of industry.' He died on 24 June 1906 and was buried in Wrexham cemetery, where 5,000 people attended his funeral.

The more private man became well known as a breeder of Shropshire sheep, and exhibited at the Royal and other agricultural shows. Prince Bismarck regularly bought sheep from Dennis. He was also a great lover of horses, and once took two of his young children, who rode ponies, on a horseback journey to Bodmin. He owned several paintings showing outstanding events in the life of the Emperor Napoleon, and it was thought by those with knowledge of New Hall that the Emperor was a hero to Henry Dennis. This was not unusual among such self-made men in the late nineteenth century. Dennis was a man of great energy, discipline and strength of purpose, who lived for his work. He was an early riser, who before breakfast took black coffee and a cigar and dealt with his correspondence. This was also the time that his works managers attended at the house to receive their orders. All of these duties had to be finished by eight-thirty, when the entire Dennis household was required to sit down for the morning meal.

Even in his last years, Henry Dennis was involved practically with the railway, although in later years it was usual for his Ruabon employees to fulfil regular tasks at and concerning the mine.

Painted portrait of Henry Dyke Dennis.
COURTESY DENNIS RUABON & J. MILNER

No job was too small for him. In the 1890s, rather than leave it to the mine managers, he settled the liability face-to-face with a local farmer whose pig had been killed on the Minsterley branch railway after getting through the SDR fence. And after the trials of *Sir Theodore* on the line in 1905, it was Dennis who as engineer told his board that a new locomotive would be needed.

Henry Dennis's sons **Henry Dyke** and **Walter Pen Dennis** were practising mining engineers.[10] Understandably they feature rather less in the history of the railway when their father was always so directly involved. H. D. Dennis appears to have become engineer to the SDR only on the death of his father. W. P. Dennis also became a director then. H. D. Dennis became chairman in June 1907 on the death of Francis B. Owen. Then it befell the brothers to carry the baton and make what they could of business at Snailbeach, until only the railway was left. Their greater task was to keep a hold on the rest of the Dennis empire.

William Edwards was also a mining engineer with Dennis & Son. He became a director of the Snailbeach Lead Mining Co. Ltd in the 1890s and of the SDRCo in 1899.[11] In the 1900s he was managing director of the Halvans Co. Ltd owned by William Kinross Arber, and invested in the Granham's Moor Quarry Co. Ltd with the Dennises in 1905. From Ruabon, Edwards later ran the Gravel Trading Co. Ltd, formed in 1928, and the Whitstone Company, both of which sold spar products from the Snailbeach spoil heaps. He lived at Bryn End, then Tynewydd and later Rhianfa, Ruabon.[12]

Other than the unrelated John Jones, three Longueville solicitors were important in the SDRCo organisation. They were **Edward Williams** (later Edward Williams Vaughan), **Martin Benson Lawford**, and **Arthur Hamilton Bardswell**. Williams replaced Joseph Whitehouse as a director in February 1886, then in August 1886 became secretary after the death of the unrelated John Jones. Lawford became secretary in 1900 when Edward Williams Vaughan retired. He became a director in 1907 when the death of F. B. Owen created a vacancy, and Arthur Bardswell took the post of secretary vacated by Lawford. It only surfaces occasionally in the Minutes that the legal advice of the secretaries to the directors was crucial to the survival of the Company. More obviously, Bardswell's letter books show that he in particular had an aptitude for practical management. The Dennis brothers supported his decisions, took no-nonsense advice from him, and relied on Bardswell to keep his finger on the pulse of the Snailbeach operation, which was only a small part of their business.[13]

H. F. Stephens – owner, director and chairman

Holman Fred Stephens' career in light railway construction and management spanned 40 years.

Portrait photograph of Holman Fred Stephens.
COLONEL STEPHENS RAILWAY MUSEUM ARCHIVE

Passport photograph of Bill Austen.
COLONEL STEPHENS RAILWAY MUSEUM ARCHIVE

He built or was associated with numerous light railways. In 1887 he studied engineering at University College London and was then apprenticed to the Locomotive Superintendent of the Metropolitan Railway. He moved to the Cranbrook and Paddock Wood Railway as resident engineer and then became an associate member of The Institution of Civil Engineers. Stephens' mission as a railway engineer was to build and operate railways at low cost, where the odds were against success. The Light Railways Act of 1896 was a boon to Stephens. His early successes included the Hundred of Manhood and Selsey Tramway in 1895 and the Rother Valley (later Kent & East Sussex) Railway in 1900. After the Great War, Stephens stayed in minor railway development, completing the difficult North Devon & Cornwall Junction Light Railway project in the 1920s. As well as his railway career he was called to the Bar, and undertook military service, mostly with the Territorial reserve, becoming a Lieutenant Colonel.

Stephens in person was a tall, striking figure, with a military bearing. Although arrogant, he had great personal charm and wit. He was much admired and liked by his staff, and held in awe for his pragmatic and innovative approach to managing the railways. After protracted illness, Stephens died on 23 October 1931.

W. H. Austen – engineer; later part-owner, director and chairman

William Henry Austen was born at Snodland in Kent on 8 May 1878. On leaving school in 1891 he was apprenticed to J. T. Firbank, contractor to the Cranbrook & Paddock Wood Railway. He then joined H. F. Stephens on the construction of the Rye & Camber Tramway, and was 'put in charge of the locomotive workshops', then becoming 'outdoor assistant' to Stephens and his expanding group of railways. He acted as resident engineer in the days of building new lines, and as a general trouble-shooter. When he succeeded to Stephens' job as head of the group in 1931, Bill Austen was known as a hard working and versatile manager, but he lacked many of Stephens' better qualities. Nevertheless he moved to ensure the continuity of the railway group, notably by purchasing 23 Salford Terrace, Tonbridge, the headquarters building. He also became director or consulting engineer to various railways within the group, including the SDR of which he became chairman in 1945. The Salford Terrace offices finally closed on 7 June 1948, after various railways in the group were nationalised and much of the consultancy work was lost. Austen, then aged 70, retired. He died on 26 February 1956.[14]

J. A. Iggulden – chief clerk; later part-owner and director

James Arthur Iggulden was born in 1899 in Tonbridge, Kent and attended St Stephen's school. In March 1914 he was employed by

Portrait photograph of Arthur Iggulden.
COLONEL STEPHENS RAILWAY MUSEUM ARCHIVE

Stephens as an apprentice in the drawing office, but soon after became involved with the book keeping and studied accountancy at college at Stephens' expense. He thereafter became chief clerk and Stephens' principal 'indoor assistant.' He was given power of attorney over Stephens' affairs shortly before the latter died in October 1931, and was one of the four beneficiaries under his will. Continuing to work at Salford Terrace until the railways were nationalised on 1 January 1948, he took the post of head of the commercial accounts section at Waterloo (Southern Region, British Railways). He acquired Bill Austen's SDRCo shares on his death, then becoming chairman. He died on 20 February 1979.[15]

Michael, Anthony and Susan Iggulden

On acquiring Bill Austen's shares and becoming chairman, Arthur Iggulden passed some of the shares to his sons Michael and Anthony, and Michael was appointed a director. Being a chartered surveyor, Michael Iggulden was made responsible for dealing with all property matters, such as maintenance of the company's land, wayleaves and the details of the leases to the Shropshire County Council (SCC). Anthony was also a director for a short time. On the death of Arthur, Michael Iggulden became chairman and his wife Susan a director. Shortly afterwards, Anthony, having no more interest in the company, sold his shares to his brother. Michael and Susan remain legal owners of the SDRCo.

H. M. Bates

Henry Montague Bates was born on 17 March 1849 and educated at Bell Burrows and Johnson's Naval Establishment, Southsea. He was Private Secretary to the Chief Secretary for Ireland for four years. In 1875 he joined the Corporation of London as an assistant clerk to the City Commissioners of Sewers and in 1895 became Clerk to the Commission. When in 1898 the Commission was amalgamated with the Corporation and became the Public Health authority for the City, he retired on a pension. He remained active, particularly in freemasonry in which he held high rank. He was also Chairman of the Eccentric Club for 25 years, doing much charitable work for Great War veterans.

As an associate of H. F. Stephens, Bates was a director of the Kent & East Sussex Railway and the Selsey Tramway as well as the SDR. He was also chairman of the North Devon & Cornwall Junction Railway built by Stephens, and a signatory to agreements made with the Southern Railway concerning Stephens' Southern Heights railway scheme. He died on 30 December 1928.

During the 1966 annual inspection, with (left to right) Cyril Hewitt, Doug Challinor and Arthur Iggulden by the level crossing at Snailbeach. Behind them is Doug's car, and Prospect Cottages are in the left background. The photo was probably taken by Michael Iggulden.
COLONEL STEPHENS RAILWAY MUSEUM ARCHIVE

J. L. White was a District Inspector of the London & North Western Railway (LNWR) at Shrewsbury, and was appointed Superintendent of Passengers and Goods on the Shropshire & Montgomeryshire Light Railway (S&MLR) on 22 February 1912. He became a director of the SDR when it was bought by Stephens. White was the local manager for the two railways.

Jeremiah MacVeagh was born in Belfast in 1870, the son of a ship owner and born into a well-known Irish Nationalist family. He began working life as a journalist, then went to London in about 1894 as the Private Secretary to Sir James Whitehead, a former Lord Mayor of London. He remained active in journalism and politics, and in 1902 was elected MP for County Down South, holding the seat until 1922. He was a witty man and won many friends in the House of Commons across all parties.

During the Great War he was active in raising the Tyneside Brigade for the British Army and in 1918 was called to the Bar. Post-War, with the Partition, he avoided provincial politics and stayed in London. After an abortive attempt to re-enter British politics as Labour candidate for Sunderland, he became successful in business. He was managing director of the Register

Publishing Company in London, and a director of various Irish companies including the Dublin & South Eastern Railway. MacVeagh was involved with many of the Stephens railways in the 1920s. He became a director of the East Kent, North Devon & Cornwall Junction, the S&MLR and Kent & East Sussex Light Railways as well as the SDR, and was a promoter of the Southern Heights Light Railway scheme. He died in Dublin on 17 April 1932 aged 62.

James Ramsay OBE began his career with the Caledonian Railway. Starting at the Perth offices, he then became a clerk for the Chief Goods Manager in Glasgow. From 1905 he was Assistant to the Goods and Plant Superintendent, becoming Superintendent in 1916. He was awarded the OBE in either 1918 or 1919. At the Grouping of railway companies he became Rolling Stock Superintendent, Northern Division, London Midland & Scottish Railway (LMSR), retiring in 1925. He was for many years active in the Railways Clearing House in London and probably became acquainted with H. F. Stephens through this connection. Ramsay became a director and subsequently chairman of the Shropshire and Montgomeryshire Light Railway, and a director of the SDR on 8 January 1931. He died on 5 February 1943.

John Pike OBE, MInstT was educated at the City of London Schools. He had a long career with the LNWR, beginning in the Minerals Office in 1884. He transferred to the Outdoor Goods Manager's Office, then to the Department of the Supt. of the Line. He served as personal assistant to the mineral traffic manager from 1902 to 1909, then as chief clerk to the general manager, Sir Frank Ree. He was appointed assistant goods manager in 1914. Pike joined the newly-formed Ministry of Transport in 1919 as director in charge of the Rates Branch, and was chairman of the Irish & English Traffic Rates Conference. He returned to the LNWR in November 1921 as assistant to the general manager (Rates and Fares). With the Grouping he served as goods commercial manager of the LMSR from 1923 to 1928, when he retired. He was co-opted as a director of the SDR on 8 January 1931, becoming chairman in 1932. He was also a director of the S&MLR, becoming chairman after the death of James Ramsay. John Pike was highly regarded among senior railway managers. Among tributes paid at his death on 2 March 1946, it was recorded that 'Pike's opinions carried weight in conference rooms where he was forthright and thorough.'[16]

Little is known of **R. F. Bowmaker** and **C. R. Hewitt**. Bowmaker was the accountant to the S&MLR as well as secretary and accountant to the SDR. He lived in Southampton. Cyril Robert Hewitt was a key member of staff at Salford Terrace, being clerk to the S&MLR, the Weston Clevedon & Portishead Railway, and the SDR to which he also became company secretary. He held the latter post until his death. On the closure of the Salford Terrace offices, Hewitt joined the Claims section at Waterloo (Southern Region, British Railways).[17]

Management at Snailbeach

The railway had its own manager from 1877 until October 1883. Railway operation then became another job for the Snailbeach mine managers, until Stephens bought the railway in 1922.

Robert Ellis – SDR manager 1877-1883
Robert Ellis and his wife Sarah Ann were both born in Denbighshire, and moved to Snailbeach for the job.[18] Ellis's salary was £90 a year, and the company paid a rent of £12 a year for their house.[19] As part of the economies made on the railway, Ellis lost the job in late 1883.

John Job – mine agent and accountant c.1858-1898
Born in Tavistock in 1832, he worked for the Snailbeach Mine Co. from at least 1858. He married while at Snailbeach.[20] In the 1860s the company built Prospect House for him to live in. He was accountant for the Snailbeach Mine Co. Ltd, Snailbeach Lead Mining Co. Ltd, Jones Mercantile Co. and the SDRCo. He died on 10 December 1898 aged 66, and is buried at Hope church.

James Coates – first captain to 1891
James Coates was born in 1825 in Appletreewick, Yorkshire. He was probably employed at Snailbeach by James Ray Eddy (see Chapter Two), since his youngest son was born in Minsterley in the 1860s. Coates was already experienced as a captain in Yorkshire lead mines. He was first captain for the Snailbeach Mine Co. Ltd and then the Snailbeach Lead Mining Co. Ltd until July 1891, when he was pensioned following a long period of illness. The mine employed several captains at any one time, each

Mine manager and local SDR manager William Oldfield walking back from the mine to his long time home at West View, Snailbeach. COLLECTION E. ROWSON

with specific responsibilities, and 'First' in this context meant 'senior'.[21]

William Oldfield – first captain 1891-1912

William Oldfield was born in 1851 at Wrexham. He was a manager at Snailbeach mine from 1889 to 1928, living at West View. He was from a mining family, and worked at Minera mine from the ages of 15 to 39. Henry Dennis brought Oldfield to Snailbeach as an underground manager and captain in 1889. In 1891 Oldfield was promoted to first captain on the retirement of James Coates, and in 1895 was appointed mine agent alongside John Job. Oldfield then managed the mine for the Halvans Co. Ltd from 1912 until his death in 1928. He also managed the SDR locally and for a while the Bog barytes mine. In about 1906 he became a partner in the Cyffty mine near Llanrwst. He was a member of the Institute of Mining Engineers and an expert mineralogist, whose services were in great demand as a consultant and lecturer. He was a local Councillor from 1895 to 1928, and was a lifelong Liberal and Congregationalist, serving as a Deacon. Oldfield's involvement with the SDRCo ceased in late 1922, when local management moved to the S&MLR office at Abbey station, Shrewsbury.[22]

Railwaymen 1877-1922

Drivers, firemen and fitters

John Davies and **James Hughes** were probably the first locomotive crew on the railway. On the January 1878 pay sheet, the earliest found, they were listed as Engineman and Stoker respectively.[23] Stoker was an early term for what more recently has become known as a fireman.[24]

John Davies was the second son of Thomas and Elizabeth Davies, the Snailbeach shopkeepers, whose shop was the origin of the name Shop Lane. In 1871 Davies was 23 years old and was a railway locomotive stoker for one of the main line railway companies. He was lodging at Old Coleham, close to the Shrewsbury locomotive depots of the LNWR and Great Western Railways.[25] Main line locomotive training meant starting as a locomotive cleaner. As well as exterior cleaning, this included clearing out the ash pan and smoke box, and lighting the fire ready for the crew starting work. In this way, John Davies had plenty of locomotive experience by the time he got the driving job on the SDR. By 1881 he had married and was already widowed, and living with his widowed mother in Snailbeach. In late 1881 he re-married to Martha Hughes, and they set up home in Little Minsterley in her uncle Thomas Whitefoot's house, where Martha had kept house for some time. They were still living there with four children in 1891. Having suffered pneumonia for ten days, John Davies died at home on 13 May 1894 aged 45.[26]

James Hughes, a former stationary engine

James Hughes in later life with his wife Hannah and family. The photograph is by his brother, Caesar Hughes, who like his brother and sister-in-law was also born near Snailbeach mine. COURTESY G. C. HUGHES

Snailbeach native James Hughes pictured c.1905 as locomotive foreman on the Glyn Valley Tramway locomotive *Sir Theodore*, wearing his customary bowler hat. The train is waiting at Glyn Ceiriog station. This was the locomotive lent by the GVT to the SDR in 1905 for trials. COLLECTION JOHN MILNER

stoker, was born on 16 May 1853 at Crowsnest, near Snailbeach, as reported on his birth certificate. He was the second son of Vincent Hughes, an engineer at Snailbeach mine, and his wife Mary. Before 1888, Hughes left the SDR to work on the conversion to steam working of the horse-drawn Glyn Valley Tramway (GVT). It is likely that this move was in 1887, when the work began.[27] (Wages records for 1886-1887 have not been found, so Hughes' departure date cannot be confirmed.) He took his wife Hannah (née Hewitt) and family with him and settled for good, being employed by the GVT as driver and rising to the rank of foreman.[28] John Davies was then joined by **Edward Wardman** as fireman. Wardman was formerly a stationary engine driver. He took the driver's job on Davies' death in May 1894, being replaced by **Edward W. Edwards** as fireman with William E. Edwards as an occasional stand-in. Edward Wardman left in June 1897, and by 1901, aged 44, was a painter's labourer living in Leicester with his wife and family.

In July 1897 Edward W. Edwards became driver, with **Benjamin Philpott**, a former lead miner, becoming fireman and William E. Edwards and **George Humphries** (one of the platelayers) occasionally standing in for Philpott. Philpott stayed until September 1901. George Humphries covered as fireman in October, after which **George Philpott Edwards** took the job until July 1902. Philpott then returned, and continued to fire for Edward Edwards until at least October 1903, when the available railway records finish. Later Benjamin Philpott took advantage of his experience and moved to Shrewsbury with his family. In 1911 he was an engine driver and living at 2 Drinkwater Street, off Frankwell.

Until 1883, a fitter **John France** was employed, virtually full time, at a very high rate of 10s a day (see Chapter Thirteen). He was previously a driver of stationary engines, which traditionally included maintaining the engine as well as driving it, so he was suited to the locomotive work. France returned to stationary engine work. After this, the locomotives were maintained by casual fitters and specialist maintenance contractors, doing the work the railwaymen could not do for themselves. In March 1894, even a fitter from the Great Western Railway was hired to repair a broken spring on *Fernhill*.

From the early 1890s various tradesmen, known by the interchangeable terms fitter and mechanic, found themselves work on the SDR locomotives. Four members of the Edwards family last mentioned among the locomotive drivers were prominent among these. They were descendants of a William Edwards, born in

Pontesbury in 1822, who was a stationary engine driver for several decades and lived in Snailbeach. The four were William's son Edward, Edward's sons William Edward and George Philpott, and another grandson Edward William the son of Edward's elder brother William junior. When the locomotive firing and driving jobs became vacant, they took those too, as described above. When necessary they could combine the work of repairing and operating the locomotives, and learnt how to drive them through repairing and testing them, as well as learning from previous drivers and firemen.

William Edwards senior introduced three of his sons, William, Edward and Ebenezer to the stationary engine driving trade. As with John France, this experience gave them the general skills of a fitter or mechanic. The eldest, William junior, was born in 1849. He married Mary Philpott (Benjamin Philpott's sister) in 1869, and in 1871 was living in Snailbeach with a young family and was a lead miner and smelter. Edward, born in 1852, still lived with his parents in 1871 and was a lead puddler, meaning he worked the lead smelting furnaces – and both brothers worked in the same place. By 1881 Edward was also married with a young family, and he and William were now both stationary engine drivers at a lead mine. The youngest of the three, Ebenezer, born in 1858, still lived with his parents in 1881 but was also a stationary engine driver at a lead mine then.

As mining expanded in the 1870s, all three brothers had been able to take advantage of the knowledge passed on by their father, but in the 1880s the slump led to a reduction in the number of working engines. William senior and junior were still driving stationary engines in 1891, but Edward was now a lead mine mechanic and using his maintenance skills more generally. He included SDR locomotive repairs among his work, and his son William E. and nephew Edward W., among others, worked with him. Ebenezer had to move away in the 1880s to stay in work, and in 1901 was still driving stationary engines in the Doncaster area of Yorkshire.

William junior's son Edward William, born in 1873, was a lead ore dresser in 1891, but as described earlier became SDR fireman and then driver. In 1901 he lived at Snailbeach mine with his wife Rhoda and younger sister Martha. He was still a railway locomotive driver in 1911, when living at the Chapel House on Lordshill Farm, and it can only be presumed that this was with the SDR. In the 1890s and until at least 1903 he did locomotive repairs, including working with his uncle Edward.

Edward's son William Edward, born in 1878, was as described earlier an occasional stand-in fireman. In 1894 he was also repairing locomotives occasionally with his father. He continued this casual locomotive firing and repair work until as late as 1897. Like his uncle Ebenezer he moved away, and in 1901 he was a railway engine stoker boarding with a family in the coal mining district of Swinton, Rotherham.

Edward's son George Philpott Edwards was born in 1879. At the time of the 1901 census, he was an underground repairer in a coal mine in South Wales, but returned to Snailbeach and fired locomotives for his cousin between November 1901 and July 1902. He occasionally repaired the locomotives until at least 1903 when the records end. Finally, Edward's son Alfred Philpott Edwards, born in 1881, was recorded assisting on a locomotive repair in 1903.

Apart from SDR work, George and one of the Edward Edwardses (probably his father, as his cousin was driving on the railway) worked together for the Halvans Co. at Snailbeach in 1906.[29] George also worked at the Bog, Cothercott and Huglith mines which are all in the vicinity. He was later employed by the Shropshire County Council at Callow Hill quarry.[30]

Platelayers, junction men and others

William Jones was the foreman platelayer, and **Samuel Jones** was the regular loader and unloader (or junction man).[31] Samuel Jones worked for the SDR until at least 1903. In 1883 and 1884, William Jones, **James Hughes** and **George Harrison** were the regular permanent way gang. It is not known if it was the same James Hughes that was the stoker.

William Jones and his family, including his son Robert who worked with him on the SDR, lived at Pontesbury between 1881 and at least 1907. (This William Jones is not to be confused with William Arthur Jones of Horsebridge, Minsterley, who joined the company later.) William moved them from north Wales to take the position of foreman platelayer. He was born in Corwen, Merioneth in about 1834, and census records show that he had moved several times with his railway work, gaining experience. His wife Jane, ten years younger, was born in Rhyl. Their son Robert was born in about 1869 in

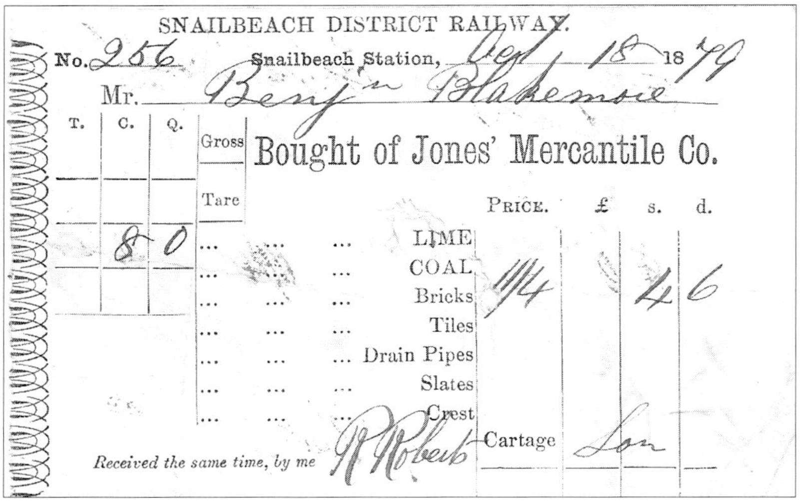

SNAILBEACH DISTRICT RAILWAY.

No. 256 Snailbeach Station, Sept 18 1879

Mr. Benjn Blakemore

T.	C.	Q.	Gross	Bought of Jones' Mercantile Co.			
			Tare	Price.	£	s.	d.
8	0			... ... LIME			
				... ... COAL	11/4	4	6
				... ... Bricks			
				... ... Tiles			
				... ... Drain Pipes			
				... ... Slates			
				... ... Crest			
				Cartage			

Received the same time, by me R Roberts

This Snailbeach wharf sales receipt issued at the Snailbeach Station weighbridge by Richard Roberts in 1879 shows the range of products which could be bought via the railway. Snailbeach mine was growing in the 1870s, and building materials were especially important for miners improving their cottages.
COURTESY M. J. SHAW

Llandderfel, near Bala, Merioneth, while Thomas was born in Corwen in about 1874. Robert and Thomas, when of working age, gave their occupations as platelayer and baker respectively. The last currently known of the Jones family is that William died in the railway's service on 5 November 1907 aged 74, and Robert still lived in Pontesbury at the time.[32]

Other track workers included **Thomas Morris** and **J. Humphreys**. In the 1890s and until 1903, the gang included William Jones, George Humphries, **Robert Jones** and **William Sankey**.[33] Sankey was a former lead miner, and like others had to find other work with the decline of mining. In 1901, declaring himself to be a railway labourer aged 56, he lived at the Poplars, Pontesford Hill, with his wife Margaret and three children. In 1911 he was still a railway labourer at the same address, and his son Charles was doing the same work.

For some years **Richard Roberts** was employed part time at Snailbeach station, weighing and selling coal, lime and other commodities. In 1881 he was a 32-year-old coal merchant's agent, as declared to the census officer. A surviving receipt issued in 1879 by the Snailbeach District Railway at Snailbeach Station records goods 'bought of Jones' Mercantile Co' and is signed by Richard Roberts.[34] Clearly Roberts understood himself to be employed by Jones' Mercantile, while the receipt shows two company names and his wages are recorded on the SDR pay sheets. By 1891 work at the Snailbeach Station weighbridge had declined, and Roberts had returned to lead mining. The Roberts family was ambitious: Richard Roberts' younger brother William, who began work as a lead miner, was a mine storekeeper in 1891 (possibly a part-time carpenter – see below) and mine clerk (for the Snailbeach Lead Mining Co. Ltd) by 1901. William's son Joseph was in later years proprietor of the Snailbeach Barytes Co., and in 1944 took the lease of Snailbeach mine (see Chapter Eight).

In spring 1883 it was decided to employ **Richard Young** as a locomotive cleaner, but this was another job which disappeared with the cost cuts later in the year. For wagon repairs, carpenters and smiths were hired. Carpenters included **John Jones**, **Thomas Preece**, **John Davies** and **Joseph Waters** in the 1880s and to 1891. In 1892 **William Roberts** (perhaps Richard Roberts' brother) first appears in records, and was used in the 1890s along with **Richard Webster** and **Oliver Oldfield**. From 1900 **William Parbutt** and Oldfield were used. Oliver Oldfield was a son of mine manager William Oldfield and born in Bersham, Denbighshire. In 1901 he was aged 20, still a carpenter and living with his parents at West View, Snailbeach. Smiths used in the 1890s included **Richard Rogers**.

In 1909 **Daniel Simister** is first known as a labourer employed by the SDR (see Chapter Sixteen). At the same time he was publican of the Horseshoe Inn, Pontesbury, where he lived with wife Fanny and three children. He had run this public house since at least 1901 and was still there in 1911, when he declared day work for the 'Snailbeach Railway Company' [*sic*] as his second job.

By the Great War the railway staff had been slimmed down. In 1914, one of the two Sankeys appears to have been working for the SDR – certainly a Sankey was. In 1916 the SDRCo declared that it had five employees: foreman platelayer, labourer, junction man, engineman and stoker.

William Arthur Jones, usually known as Will or Bill, was a lead mine labourer in 1911, probably at Snailbeach. He lived at 4 Horsebridge, Minsterley with his wife and two daughters.[35] Will later became the SDR junction man, and was certainly with the company during the Great War. In 1916 he was described in a letter to the Ministry of Munitions as 40 years old, living at 9 The Hollies, Horsebridge, Minsterley, '... *indispensable to the Railway* ...' and '... *the only man in charge of all the traffic* ...' He survived the transfer to the Stephens railway group, and retired from the railway on 3 July 1946 having worked to age 70 '... to get the extra half crown on his pension.'[36]

No.2 with Will Jones, left, and Dickie Preece on the footplate at Pontesbury, early 1940s. Will Jones was Betty Hordley's grandfather.
L. HORDLEY, COLLECTION K. C. LOCK

RAILWAYMEN POST-DECEMBER 1922

Junction man
From 1923 **Will Jones**' daily routine included taking full wagons down to Pontesbury by gravity, so it was easier to start his working day at Snailbeach. He walked every morning from his later home at 4 Horsebridge through Minsterley and up the Callow Lane, then along the railway from the skew bridge. Reaching Molly Wardman's shop a little after 9 a.m., he dropped in for a packet of cigarettes. His next port of call was Gray's spar gravel works, to take any full wagons down to Pontesbury.[37] Will kept records of all incoming and outgoing goods and minerals, took the orders for wagons, and ensured that minerals were deposited in the correct standard gauge wagons for the consignees. He dealt with timesheets and wages and customer payments; and passed on all the instructions borne by the letter post from Tonbridge. The Tonbridge headquarters was kept informed of all traffic, usually in writing, for which standard forms were used. Judging by the comment to the Ministry of Munitions (see above), his duties before the Stephens takeover were similar.

Fitters and drivers
Little is known of the fitter Mr **J. Webb** other than that he stayed at 14 Counting House, Snailbeach for a little while, and worked at Pontesbury sidings on the re-gauging of the Baldwin locomotives in 1923. It is not known if he drove the locomotives. By 1924, **William ('Bill') Crawley** was fitter-driver on the railway. He was not a local man, but lived with his family at 4 Farm Cottage, Snailbeach until about 1935.[38]

Fitter-driver **Thomas James Gatford** moved to the SDR from the Bishop's Castle Railway according to Eric Tonks, but this is highly unlikely (see below). Tom Gatford worked on the SDR from as early as 1929 and perhaps until 1936. Gordon Cooke knew him and was asked to

Tom Gatford, former SDR fitter-driver, on a Terrier-class locomotive of the Weston Clevedon & Portishead Railway. Gatford's boots always shone like mirrors, giving away his military background.
P. STRANGE

Fitter-driver David Reese Jones on board *No.4* at Pontesbury, c.1936. The locomotive is on the west siding and has a Bagnall wagon behind it. The practical results of the gauge widening of this locomotive can be seen – there is little if any overlap of the tyres on the rails, which gave no problems at the usual slow running speeds. The re-railing jack on the front footplate was essential on all the SDR locomotives, to put them back on the road in the event that a wrought iron rail head sheared through.
J. F. BRUTON, NATIONAL RAILWAY MUSEUM/SCIENCE & SOCIETY PICTURE LIBRARY

stand in as driver once when Gatford had 'flu. This confirms that Gatford was working on the railway after 1934, when Gordon started work as a quarryman at Callow Hill. At some time Gatford lodged at the Counting House and in the 1930s lived at Farm Cottage. The S&MLR and the SDR were both managed from Abbey Station, Shrewsbury, and staff were transferred between the two lines as work dictated, with this happening to Gatford at least once.[39] Then in 1936 he was moved, supposedly from the S&MLR to the Weston Clevedon & Portishead Railway (or WC&PR – another of H. F. Stephens' railways) to become senior driver. He worked on that line until it closed, and drove its last steam-hauled train on 18 May 1940.[40] It seems that Tom Gatford's promotion actually left a vacancy on the Snailbeach line, and a string of fitter-drivers who have all been identified followed him in the period 1936 to 1947.

Eric Tonks reported that Tom Gatford drove the tractor on the SDR for a while before retiring, but the driver who did this was George Edwards (see below). A photograph in Eric Tonks' book also purports to show Gatford driving an SDR locomotive when he was in fact working on the WC&PR at the time. Genealogical and other research, and the discovery of a crucial memo written by Bill Austen, has uncovered Gatford's actual life during the Second World War as well as giving supporting evidence of his earlier background.[41] Although not strictly part of the history of the SDR, it is worth recounting the findings to reinforce the other evidence, which already overturns the previous account.

Thomas James Gatford was born on 22 January 1874 to James, a general labourer, and Jane, of Bookhurst, Crawley, Surrey. At age 31 he married Daisy Ethel Wood, aged 26, on 1 July 1905 at the registry office in Strood, Kent. At this time Tom Gatford was a corporal in the Royal

Jack Owen on board *No.4* at Snailbeach locomotive shed, c.1939.
COURTESY INDUSTRIAL LOCOMOTIVE SOCIETY

Engineers based at Lydd in Kent. His father was described as an engine driver, and Daisy, the daughter of a barge captain, lived at Frindsbury in Strood. While serving on the WC&PR, Gatford said that he was a former Regimental Sergeant Major. The civil record confirms Tom Gatford's military service, and gives a good indication of possible contact with H. F. Stephens, leading to his later employment on Stephens' lines. Stephens is well known for employing ex-servicemen he knew or those who served under officers he knew. When interviewed by Selwyn H. P. Higgins, no former Bishop's Castle Railway staff recalled Gatford, and it seems certain that he did not work on that line.

When the WC&PR closed down, Tom Gatford was aged 66, and past normal retiring age. The railway company helped the Gatfords with a gift of three guineas to move to Rochester, Kent. Here Gatford got a job as an aircraft works policeman (what would now be called a security guard), and was no doubt living near family and old friends. He died of chronic bronchitis aged 72 on 15 February 1946 at home, 17 Foord Street, Rochester, Kent. In the unlikely event that he ever returned to Snailbeach during the war, it will have been very briefly, and only as a visitor, not a locomotive driver. All witness and documentary evidence supports this view.

David Reese Jones was a fitter-driver who worked on the railway in about 1936-38, but had to give up the job because the heat from the locomotives aggravated a skin complaint.[42] He was a Baptist lay preacher, and lived up on the Stiperstones at the Paddock Farm.[43]

Fitter-driver Dodd (or Gostlow) on *No.4*, shunting at Callow Hill sidings.
PETER CLAY, COURTESY LENS OF SUTTON

John Henry ('Jack') Owen was born in Snailbeach in 1893. He took the job of driver (whether he was also a fitter is unknown) around the beginning of the Second World War, but again left after a short time, to be a stoker at the Minsterley Creameries where he stayed until his retirement.[44]

A. L. Dodd and **R. Gostlow** appear to have come and gone in quick succession between 1939 and 1940. Dodd appears to have been transferred from Kinnerley. An ex-LNWR fitter named Albert Dodd worked there. Gostlow stayed at 5 Snailbeach, the Farm House, which had been divided into flats. He was not a local man, and moved on quickly for better money.

Richard ('Dickie') Preece followed the last three named onto the footplate in the early 1940s. His earlier experience was as a stoker at

Pontesbury in the early 1940s. *No.2* with a train of seven empty wagons including one of the Bagnalls in the centre. The men are possibly (left to right) Will Jones and Dickie Preece. The point leads to the tarring plant.
H. F. G. DALSTON

No.2 with Dickie Preece at Pontesbury, early 1940s. Dickie was Gordon Cooke's uncle. Note the shunting rope stowed around the coal bunker.
L. HORDLEY, COLLECTION K. C. LOCK

Snailbeach mine, and later driving stationary engines.[45] He left the railway in November 1945.

Replacing Dickie Preece, George P. Edwards returned to the railway as fitter-driver on secondment from the SCC, and after the locomotive boilers were condemned he drove the tractor on the railway. Although George was offered a permanent position with the SDR in 1945, he never left the employ of the SCC until his retirement. His working on the railway was a convenient arrangement made before the leasing of the railway to the SCC, and the SDRCo accounts covering George's time on the railway show wages paid for locomotive work. George took special instructions about locomotive maintenance from the SDRCo via the SCC.[46] It can hardly have been easy to find a new locomotive driver to work in the wilds of the Shropshire hills in 1945 – Bill Austen used George on the recommendation of Will Jones. The Minutes of the SCC Roads & Bridges Finance Committee record that a temporary agreement was made whereby the Council could operate its own traffic using the tractor on the SDRCo's track (see Chapter Nine) – and George was the operator. Driving the tractor up and down between Pontesbury and Callow Hill a few times a week was scarcely a full-time job, and was probably something George did between maintenance tasks at the quarry. He died on 21 March 1956, aged 77.

Platelayers

The first one known by name is **Edmund Lewis**, who took the job in the 1920s.[47] A Snailbeach native, he was a blacksmith (as was his father William before him), and had a smithy at Lower Works, Snailbeach. The decline in mining reduced the work available for smiths, and also the local trade in horse-shoeing and general work. So Edmund turned his skills in another direction and joined the SDRCo. Like the other railwaymen he was expected to be versatile, and he doubled as a brakesman.[48] Edmund died on 15 February 1949 aged eighty and was buried at the Methodist Chapel, Snailbeach.

Jackie Belter (John Arthur Rowson).
COLLECTION J. BUTLER.

Jackie Belter (John Arthur Rowson) enjoying his Sunday rest, with a good book, gramophone records and a friend for company. COLLECTION J. BUTLER.

Before leaving the railway, Edmund taught **Jack Hughes**, who did the job in the 1930s. Jack cycled to work from Hanwood every day. In mid-1933 an assistant to the platelayer was employed for a few weeks at 30s a week. In a memo dated 19 October 1940, Will Jones was officially informed of Jack's resignation, and that his replacement was to be paid 48s per 48 hour week. Jack's successor was **John Arthur Rowson**, a Snailbeach native usually known as Jackie Belter or Jackie Rowson. Previously he had done surface work at Snailbeach mine such as mineral dressing. On the retirement of Will Jones in 1946, Jackie took on Will's remaining duties. With the lease of the railway to the SCC in 1947, Jackie transferred to the latter and carried on working on the line. He died on 22 February 1975 aged 78 and was buried at the Methodist Chapel, Snailbeach.

No.3 hauling empty wagons in the cutting en route to Callow Hill with railway enthusiasts on board. Note the steel sheet hopper lining: the dome-headed coach bolts holding the cladding can be seen. J. G. VINCENT

CHAPTER SIXTEEN

Incidents, accidents and anecdotes

Recollections and records of events of various kinds find a home in this chapter. They include the serious and the sobering, as well as the everyday things which nobody thought twice about when the railway was among them. When combined with knowing its general workings, they help to form a picture of life in, on and around the railway, and so put it into perspective.

ACCIDENTS

Injuries to life and limb were rare among the Snailbeach railwaymen. Only one fatality and one injury on the SDR were recorded in the Board of Trade returns, which cover the period up to the Great War and a little after. From the early 1920s, the new Ministry of Transport only issued reports on the most serious accidents, issuing summary booklets of statistics for the rest. If there were other reportable accidents on the SDR, they cannot be identified from the latter source. Local residents have only recalled an injury to Will Jones, so it seems unlikely there were any serious accidents after 1919 when this happened.[1] Only one other recorded accident on the line has been found. This concerned the death of a member of the public using the railway as a short cut. No doubt there were other minor accidents to railwaymen, and there were certainly accidents to others.

DERAILMENTS

As described earlier, wagons are a significant part of a railway, and derailments can have a critical effect. Memos by W. H. Austen sent from Tonbridge to Will Jones help to illustrate this:

> 2nd March, 1939
> Derailment of Wagon No.8. Broken Axle 27.2.39
>
> I note your letter of the 28th ultimo. Do you see that the wagons are well oiled? If you happen to get a hot axle on one of these wagons, the axle will easily break. Please let me know.
> I presume the spar was transferred into another wagon and that we shall not get a claim for lost or dirty spar?

The following memo seems to refer to a derailment late in 1945 mentioned in Chapter Eight:

> 12th April 1946
> ... can you let me know what materials will be required to repair say six wagons to go on with. Rowson says he has cleared up the wreckage of the six wagons which were smashed up some time ago, as best he can ... suggests borrowing hawser and use engine to pull wheels up the embankment.

FATAL AND SERIOUS INJURIES TO RAILWAYMEN

The fatality was the result of an accident on 16 November 1901, where the official return recorded one person 'killed whilst moving goods/luggage in stations/sheds (not working crane) – staff or contractor.'[2] This does not quite fit the circumstances recorded at the Coroner's inquest on 25 November by the *Shrewsbury Chronicle*. Of the available categories of accident for the Board of Trade statistics, perhaps it fitted this one best. The newspaper report published on 29 November 1901 leaves little unsaid.

> FATALITY AT SNAILBEACH – Mr R E Clarke (coroner) on Monday evening held an inquest at Shrewsbury on the body of Edward Perkin, aged 50, of Plox Green, near Minsterley, a sawyer in the

Ex-Granham's Moor hopper wagons, showing the sprung draw gear, loose coupling (a potential hazard for railwaymen) and strip steel components including the dumb buffers. Coach bolts hold the timber cladding to the steel hopper lining using external nuts. C. PEALLING

employ of the Snailbeach Mining Company, who died in the Salop Infirmary on the previous day. Mr Oldfield, captain of the Snailbeach Mine, was in attendance.– William Henry Perkin, deceased's son, who is an engine driver in the employ of the Minsterley Barytes Company, said his father was in his usual health when he left home for work about 6-30 on November 16th. On the 18th deceased told him he had met with an accident. Witness took him to the Infirmary on the following day by order of a doctor who had been called in. Deceased had been in the employ of the Mining Company for four years.– George Henry Humphreys, Pontesbury, a labourer in the employ of the Snailbeach Mining Company, said he met deceased at Pontesbury Junction on November 16th. There were eleven or twelve men altogether, and they were engaged in removing a quantity of crushing machinery from Pontesbury up to the mines. They left Pontesbury about 11 o'clock. When the train got within a short distance of the spot where the machinery was to be placed a portion of the machinery slipped from its position on a truck and crushed deceased's leg against a piece of timber. There was no jerk of the train when the accident happened. When the deceased extricated himself there were two small bruises on his left knee. Deceased continued to work throughout the day, and did not complain of being seriously hurt.– Dr C P Burd, house surgeon at the Salop Infirmary, said that when deceased was admitted to that institution on the 19th inst. he was suffering from stoppage of the bowels. An unsuccessful operation was performed on the 23rd inst. Deceased died on Sunday morning. A post mortem examination showed that deceased was a strong, healthy man. It was possible for the injuries to have occurred through a blow.– The Coroner: Had the man been sent to the Infirmary earlier could anything more have been done to save his life?– Witness: I don't think so.– Several members of the jury stated that sufficient evidence had not been produced to prove that deceased died from natural causes or the injuries he had received.– The Coroner said there was not the slightest doubt that deceased's death was the result of the injuries. However, if the jury wished he would adjourn the inquiry for the production of further evidence.– After some desultory conversation the jury returned a verdict of "Accidental death."

Although he thought little of it at the time, it appears that Edward Perkin caught a serious blow to his abdomen as the machinery shifted on the moving train, as well as being trapped by the leg to which the only visible injury occurred. It was only the obvious trapping of his leg that witnesses reported. It only remains to note that he was considered a contractor to the railway: the men were hired casually for moving the machinery (see Chapter Twelve).

The serious reported injury occurred on 17 August 1909, when labourer Daniel Simister had an accident *'while braking, spragging or chocking wheels.'* The case was referred for arbitration under the Workmen's Compensation Act. The hearing at Shrewsbury County Court on 3 October 1910 was not reported in the *Shrewsbury Chronicle*, although other hearings were.[3] Shunting the wagons gave junction man Will Jones a permanent injury following an accident on 25 April 1919. His left thumb became locked against his palm after his '... *left hand was badly crushed whilst coupling waggons.'* He spent some time in the Royal Salop Infirmary and was paid damages from insurance under the Workmen's Compensation Act.

Although strictly a mining accident (and it was reported as such), it is also worth noting a death which happened while shunting in the Snailbeach mine sidings. On 5 August 1897, Mr W. Edwards, employed as a 'locoshunter', was run over and crushed by a flat wagon being pushed by a locomotive. He fell off in front of the wagon while stooping to lift a coupling chain.[4]

PONTESBURY

Since the privatisation in the 1990s of Britain's railways, passing the buck between companies has become well-known after accidents, to deflect liability for damages claims. This is nothing new. In May 1897, the Joint Railway engineer's office reported to Dennis & Son a derailment of wagons on the standard gauge coal siding at Pontesbury. The Joint Railway engineer's inspector had told him that the lines to the turntable-tippler were out of order, and that the turntable-tippler was unsafe to work over with engines. Dennis & Son asked William Oldfield to investigate, and he replied on 28 May:

> William Jones says that he cannot possibly do anything more to the line than he has done, and that when ... Mr. Johnson's inspector was up last week, he tried it with a gauge, a staff and level, and said it was alright and no fault to be found. ... The Engine does not go over the table, and the day the wagons went of the rails it was entirely the [*branch*

> *line*] Railway Companys men fault in coming in too fast and sending them down to the end of the line, and then drawing them back with a rope on an angle ... [*sic*] [5]

The assumed ignorance of the inspector is perhaps inexcusable, but looking from the other side, a lack of understanding by customers of the facilities at Pontesbury could lead to delay and frustration. An example can be found in correspondence of 1907 with the Granham's Moor Quarry Co. This company had ordered two wagon loads of coal, and complained that they had stood at Pontesbury for over a week without being unloaded. After investigating, Arthur Bardswell wrote to the Granham's Moor Quarry Co. that '... *our men are unloading the wagons as fast as they can but they complain that your wagons have side doors & they have to be unloaded with shovels ... If you would send wagons with end doors ... the wagons could then be passed onto our tippler & unloaded at once.*'

THE LOCAL PET – USUALLY HARMLESS, SOMETIMES OVERFAMILIAR

The SDR was very much a 'pet' railway, and as much a part of the community as Hughes's smithy, the Post Office or Molly Wardman's shop. Being the quickest way to Pontesbury, it is not surprising that the track was used as a footpath. People walked the line countless times without mishap, but on one known occasion the short cut gave fatal results. The *Shrewsbury Chronicle* of 23 January 1885 reported the death of Margaret Hampson, step mother of Richard Hampson who was a miner at Pennerley. She went from Pontesbury to visit him on 14 January intending to be back before dark, using the Snailbeach railway as a short cut. She was found dead in the sidings at Pontesbury the following morning. She had apparently fallen ten feet, there being no suggestion of any train involved.[6]

It was quite easy to get a ride on the line. The wives of mine captains, such as Coates and Oldfield, were privileged, and often organised shopping trips to Pontesbury by railway wagon.

8 October 1939. Looking north from the crossover at the Snailbeach road to rail wharf. J. G. VINCENT

They would sit in an open wagon – more likely the miners' car – and ride down with braking attended to by one of their husbands' men. We are not told how they got back to Snailbeach.[7] At other times young men from the Pontesbury area, when attending a dance at Snailbeach, would risk their necks and take a wagon down late at night to save a walk. When the engine was in steam, a lift down to Pontesbury on the footplate was not unknown either.[8]

Like the other railwaymen, platelayer Edmund Lewis was expected to double up on other jobs. One of these extra duties was as brakesman on the minerals. This meant that young Nancy Lewis and her friends could be treated to a ride down the line to go blackberrying. With the children riding on the wagons, father would bring the run to a halt where there were good pickings on the lineside hedgerows.[9]

The little railway, not being too busy, was a playground for the youngsters like almost any other place out-of-doors. They had to make their own entertainment. During the Great War timber drive, tree trunks were stacked at Top Wharf to be loaded by crane onto the wagons. In 1917, nine-years-old Molly Rowson suffered a broken leg and younger sister Nell a sprained ankle when one of the stacked trunks slipped and fell on them.[10] Wagons standing at Snailbeach were fair game at play time. Mary Challinor recalled that it was great fun to let the side of a wagon down, but it was usually too heavy to put back up again. More of an adventure for Raymond Rowson was to ride a wagon on the brake from the wharf down to the Halvans bridge – but no further.[11] Ruth Chesters remembered: *'The greatest thing happened coming home from the Stiperstones school. We would run all the way to see if we could get to the end of the railway line to have a ride on the engine coming across the back to the engine shed. Those were very happy days.'*[12]

Good neighbours

Quarrymen at Callow Hill could rely on junctionman Will Jones for a favour. The quarry was a two mile walk from Snailbeach – unless you fixed it with Will. A quick telephone call to Will and a bogie wagon would be put on the previous up empty working, and left at Snailbeach for the men to ride down in. The bogies were ideal, because the upright screw brake was easier to use than a side lever.[13] These wagons were also sprung, but whether this really made the ride more comfortable is doubtful!

Likewise, the teamwork involved in keeping the railway running extended outside the usual payroll. The Callow Hill quarrymen were a source of help if it was especially needed. Once, when Tom Gatford was struck down with 'flu, he asked young Gordon Cooke to stand in as driver. Obviously this needed the quarry manager's permission, but to say no would have meant no empty wagons to keep the supply of stone

An unidentified bogie wagon, probably No.56. The two bogies have different patterns of wheel. At some time, bogies or wheels from 'Modell 18' wagons have been swapped over. The workman at the brake standard has not been identified. The wagon is standing on the tarring plant siding at Pontesbury, and the structure just behind is part of the plant.
H. F. G. DALSTON

The fitter-driver, probably Jack Owen, stages a light engine working for a visiting enthusiast. J. G. VINCENT

flowing. Gordon also remembered helping Jack Hughes repair the track. As Gordon said, *'You had to help somebody!'*

As well as bringing mutual benefit, this co-operation fitted in with the good-neighbourliness prevailing in the district. And it certainly pays to keep in with your neighbours. On wash day, the fitter-drivers took care to warn the residents at the top of the village when the locomotive was about to come off-shed. The Snailbeach ladies could then take their washing in before the engine passed by throwing out thick black smoke.

OTHER ANECDOTES

The railway was traditionally used to bring the miners' wages up to Snailbeach once a month. In the early 1930s, Gordon Cooke worked in the spoil heaps washing spar for a little while before getting the job at Callow Hill quarry. The wages train ran even then: *'There was a carriage especially for the bloke to come up on and they used to bring the wages up once a month. They would pay us out there. I used to get two pounds one month and one pound ten shillings another for picking spar, fifty shillings a ton.'* Earlier, and depending on the employing company, the wages for miners and railwaymen were paid out in the mine office. It is uncertain what carriage it was that Gordon mentioned. From available records it seems unlikely that the miners' car survived even as late as the Great War.

Young railway enthusiast Graham Vincent visited the railway several times during the Second World War. It was reported to him that the workmen's car in its day covered the journey from Snailbeach to Pontesbury in 4 minutes at a speed of 50 mph. This is perhaps one of those tales which has improved with the telling.[14] Another later visitor, Vic Bradley, said that the wagon he rode on down to Pontesbury travelled *'like the wind'*.[15]

As related in Chapter Seven, the borrowed Glyn Valley Tramway locomotive damaged a house at Snailbeach when it came off the rails. The house was the Hewitts' old place, by Black Tom shaft. Even the new locomotive *Dennis* was not problem free. It was reported to Ken Lock that *Dennis* would not quite pass under the Halvans bridge when it was delivered. Some hard labour was needed to dig out and lower the track a little. The delivery of one or other of the large locomotives, *Fernhill* and *Dennis*, led to

3 September 1955. A rare north view of the Halvans bridge. A. J. PRATT

the doors of the locomotive shed being modified. The left-hand door post was moved to a new position bolted to the outside face of the wall, to give six or more inches extra clearance. The left-hand door was widened to suit. The evidence of the post's original position could still be seen in the 1980s.

The few stories collected here are how the tellers chose to remember the railway. To most of them, the working of it was by-the-by. The Snailbeach line was a sleepy little backwater of a railway, never attracting much attention from the outside world, and it could within reason carry on as it liked. In that sort of atmosphere, carefree and informal events like the ones described tended to flourish happily, giving the line its own vivid character. The local people living and working on and alongside the railway made it their own by using it that way. It was a 'pet' railway in the best sense of the word. But having a job to do, it had its serious side, and this is shown by the reported accidents. The accidents remind us that like any other railway, it had its dangers. It was dirty but could be a delight. Molly Rowson seemed to sum up how people felt about it when she said: *'We loved the little railway, dear, and we were sorry to see it go.'*

CHAPTER SEVENTEEN

The Snailbeach mine and smelt works

As with the engineering and operation of the railway, it would be rather useless to leave the description of the railway in isolation, when the main reason for the railway's existence was to serve the mine and the smelt works. This chapter provides a brief introduction to how a metal mine is worked, how the metallic minerals are formed and the metals extracted, and what uses these metals are put to, but focused especially on Snailbeach.

While describing the new transport at Snailbeach introduced by Henry Dennis in the 1870s, it is also useful to look at changes at the mine which he made at about the same time. As with transport, the change in contract conditions for the mine manager when Dennis replaced James Ray Eddy (see Chapter Two) was a catalyst for shrewd new capital expenditure which was long overdue. These important new mining works by Dennis are also described here.

For decades before Dennis became manager, there was a great deal of mechanisation already in use for ore dressing, and Eddy & Son had in their time greatly improved the profits made. Once again, the methods lacked sufficient economy, for a number of reasons. They either used too much water or fuel, or used too much labour (either men or horses). As with less efficient transport, such enhanced costs could be absorbed while the selling price of lead remained high, but not by the time Dennis joined the mine company.

Labour

With the exception of management and supervision, underground jobs were filled on short term contracts at the monthly 'lettings', where rival teams bid against each other to secure the work. The head of each partnership was paid in cash, and the partners then received their share of the money.[1] With underground work being difficult to control, it was important for the company to use payment methods which tied productivity to the price paid. Forms of piecework were preferred to payment of hourly or daily wages. Tributers received a proportion of the market value of the dressed ore they raised. Tutworkers sank the shafts and drove the levels – the exploration – which exposed the ore and made it accessible. This was the most easily recognisable piecework, where the men were paid by the yard of driving or sinking (or at mines other than Snailbeach, by the fathom). The name came from the German *todwerk*, meaning dead (unprofitable) work. A proportion of the mine's costs was put to such exploration, with the hope of finding new mineral reserves. Skilled tradesmen, such as smiths, carpenters, stonemasons and fitters, were paid the going daily rates, but in instances such as the drill sharpening smiths they might be paid piece work. Unskilled labourers, doing many kinds of work, might be paid wages or piece rates.[2]

Above ground

Water supply

Whether powered by manual labour, horses, steam engines or waterwheels, ore dressing was always thirsty for water. Snailbeach is high on the Stiperstones ridge, and so has a small catchment area. A pool fed by a stream was all that was available in about 1850. Dennis's large new reservoir was so important that he began surveying for it in 1871, within months of his appointment. A Dennis survey of 1872 shows it complete and in 1876 Dennis reported that *'The Reservoir has served us in good stead.'* There were also smaller reservoirs built. In 1863, James Ray Eddy planned dressing floors, a reservoir and a conduit to serve Chapel Shaft (Earl Tankerville's Vessons mine), so it is possible that these reservoirs are Eddy's.[3] They were able to feed the main reservoir by a system of pipes and siphons.[4] A buried pipe and siphon system was also used for feeding the various machines on the dressing floors below the main reservoir. This worked because the depth of water in the reservoir (and the height difference between its surface and the machines) created pressure. The locomotive shed water supply was also fed by this system. The source of incoming water for the reservoir is not obvious, although it clearly has surface water catchment. A stream is shown feeding the old pool from the east in a Minsterley estate map of 1766.[5]

Dennis's reservoir, looking across the dam in the mid-1990s. D. A. BREWER

Ore dressing processes

Ore dressing is a catch-all term for the preparation of the ores for smelting or sale. The dressed ore was known as 'concentrates'. At Snailbeach, any amount of vein material brought to the surface could contain lead ore, zinc ore, barium ores (barite or witherite) and calcite (spar). The miners were targeting one mineral, but the others could be found surrounding it – known as the gangue. The first object of ore dressing was to separate these minerals from each other and remove waste. The second, especially in the case of the lead ore, was to reduce the particle size which made it better for smelting. Small particle size also allowed easier separation in the various flotation dressing processes. The roller crusher was therefore a very important machine.

The hand-picking table was the only stage of ore dressing which could not be mechanised. It required a trained eye to pick out oversize lumps which could be fed to the crusher after bucking (a bucker is a large flat hammer for breaking or bucking lump ore), and pieces of waste which were best broken off and thrown away. The various screens separated the material by size using gravity. After a visit to Snailbeach in 1860, the French mining engineer Moissenet wrote: *'Because of the need for a nearly pure concentrate, breaking and sorting by hand can only be used to separate lumps of waste, all the rest is sent to the crushing cylinders. ... Because the galena is not finely disseminated, and the gangue is soft, however, it can be crushed into lumps of over ³/₈ths diameter before being fed through the cylinders.'* Moissenet's paper has been ably translated into English and commented on by Hazel Martell and Mike Gill.[6]

The remaining stages – buddles, jiggers and the like – were all machine operations which separated the minerals according to the differing weight of similar sized grains. They could be powered in various ways and needed water. Horse power was the most expensive power source and became increasingly expensive in the nineteenth century as the price of feedstuffs rose. Horses could drive almost any machine by being walked or trotted in a circle. This rotation could then be speeded up or down and transmitted to another drive shaft using gear wheels, and converted to linear motion using for example a connecting rod or a cam device. Steam engines needed greater capital investment than horse engines but were cheaper to run for the same amount of work, and could deliver greater force. They produced linear motion which could be converted to rotary motion, and again which could be geared and sent in another direction. Water engines, such as waterwheels or water pressure engines (the latter was unknown at Snailbeach), were generally priced between the two extremes, but were virtually free of operating cost provided sufficient water could be collected above the dressing floor. Manual labour was perhaps the cheapest, and could move a lever or turn a handle, but could only make a relatively small and intermittent effort and so was only suited to working fairly small machines like a small jigger or a dolly tub.

So we can see why Dennis completed his reservoir project so quickly. Water supply for ore dressing needed improvement, but when available in sufficient quantity could also be used to power machines 'for nothing'. Waterwheels were in use on the Snailbeach dressing floors before the Dennis era, but now were less likely to fall out of use during drought, although frost remained a problem. The No.2 dressing floors were equipped with two round buddles, driven by a six feet diameter waterwheel. They continued working until at least the mid-1880s. The No.1 dressing floors also had a waterwheel driving two buddles.[7]

Coal, boilers and steam power

The next most cost effective power source was the steam engine. Although it needed coal, which needed to be bought and transported, it could provide the power to drive any of the dressing processes – and do any other job – at any time of year. In 1872 there were three steam engines on Snailbeach mine – one each for pumping water and winding ore and rock from underground, and one for the crushing machinery which also drove the carpenter's saw mill.[8] In 1875 Dennis rebuilt the crushing engine, and adapted it so that it could also drive the various jiggers and buddles on the No.1 dressing floors using line shafts. The first six jiggers were mechanised in 1876. The engine was an 'all-indoors' type, with a cylinder of 20in diameter and four feet stroke, and the beam mounted on a cast iron frame. It had two flywheels. The crusher house was in two stages. The first pair of crushing rolls were each 2ft 4in diameter and 18in long, with tumbling shafts, a screen and a raff wheel. The second pair of rolls were each 20in diameter and 15in long. There was an elevator. Originally the crushing engine had its own boiler. Later a central boiler house was built containing three Cornish boilers, which served the crushing engine, the saw mill and a pair of air compressing engines. These boilers were fed water by a donkey engine. The boiler house and compressor complex was completed in 1881.[9] Three egg-end boilers

The mine on 22 March 1966, looking up to the summit of Resting Hill. The old count house stands above and behind the dilapidated crusher complex to the left, and the compressor house and central boiler house chimney dated 1881. On the skyline to the right is the Cornish engine house of 1848, probably built by Capt. Jonathan Harrison. P. G. HINDLEY

An early postcard. The twin miners' cage headgear over George's Shaft, with the tramway on trestles to the crusher house and the mine office just visible in the background. This was changed to a single cage headgear and winding engine by the Halvans Co. COURTESY IRONBRIDGE GORGE MUSEUM TRUST

outside the compressor house served as air receivers, and a pipe ran from them down Old Shaft to feed the compressed air-driven rock drills.[10] At the end of 1891 it was agreed to install an air-driven winch at the 342 yard level, to wind ore up a railway incline from the lower levels.[11] It was also used unofficially for man-riding, and eventually led to a fatality.[12]

The coal yard was directly beside and above the new boiler house, and served by a siding from the SDR, so that coal was delivered with the minimum of effort. The cage winding engine had its own boiler, and this installation was completed in 1872. The cage transport system, to take the miners down to the 252 yards level, was complete and working in 1873. According to a plan of the mine dated circa 1873 and the 1881 Ordnance Survey, the No.3 dressing floors were built between these dates, and were powered by a vertical steam engine with its own boiler. This had a cylinder of 9½in bore and 18in stroke and a five feet diameter flywheel. It drove a crushing machine with rolls 15in diameter and 15in long and a twelve feet diameter raff wheel, and three double-compartmented jiggers via line shafting and belts. In 1890, in locations unknown, another new crushing machine was built and the dressing machinery improved.[13]

The pumping and winding steam engines were on the side of Resting Hill, above the mine. This complex, together with the New Engine shaft and haulage level (known then as the landing level but today as the Day Level), appears to have been completed in 1848 by the mine captain Jonathan Harrison, and was fundamental to later deep mining.[14] Dennis installed an inclined railway siding for winching coal and other supplies up to these engines, where it had previously been hauled up on carts. The pumping engine was of the Cornish type, and had a cylinder 60in diameter with a stroke of ten feet. The winding engine in the 1880s had a vertical 22in diameter cylinder with a four feet stroke, mounted on a cast iron frame with a cast beam. It had a twelve feet diameter flywheel and wound the kibbles to and from the 342 yard level using flat ropes. The Cornish engine had its own pair of egg-end boilers. The pair of steam winch engines and the winding engine were served by a further pair.

In the late 1880s, with the new proprietors the Snailbeach Lead Mining Co. Ltd working under extreme economy and with limited output, Dennis reduced the amount of steam-operated machinery. A note of boiler insurance shows that the steam engine on the No.3 floors was no longer working. Neither was the winder at Chapel shaft, and the engines on Resting Hill were powered by only one pair of boilers.[15]

Transport

Serving the mine and smelt works by sidings off the SDR drastically reduced the cost of horse work. Previously coal, timber, machinery, tools and other supplies were all carted up the hill from Minsterley station. The tonnage of coal was several times that of lead going in the opposite direction. Eddy's horse tramway was an improvement on carting between the mine and the smelt works – a very short distance – but brought nothing on the return journey. Dennis's railway brought traffic in both directions and completely eliminated horse work outside the mine; and almost every important point on the mine was served by a siding.

BELOW GROUND

Old Shaft at Snailbeach was in early days a shaft for pumping and winding, but when the New Engine shaft became operational (probably in 1848) it was converted to a ladder shaft. Old Shaft was the main entrance to the mine for the miners, who in 1863, out of an eight hour shift, spent one and a half hours walking and climbing to and from the stopes. By contrast, the shift for surface workers was twelve hours, with one and a half hours allowed for meals. By 1873, Dennis had deepened Old Shaft from 192 yards to 252 yards deep, and installed a pair of miners' cages (or gigs) wound by a steam engine to ease this part of the journey and improve productivity. At its peak, the mine was worked underground 24 hours a day in eight hour shifts.

Exploration and mining

The bed rock within which the ore bodies are found at Snailbeach is hard, and difficult to work. Consequently Snailbeach was relatively expensive to mine, leaving aside all other factors. The use of compressed air-driven rock drills began early in 1882. There were four Winn's rock drills and six by Sandycroft. These greatly reduced the time and labour involved in breaking rock. Previously all holes for explosive charges were drilled by hand. The hand drills were long steel rods with a blade forged on the end. The first man held the drill in position, while the second man hammered its blunt end. With each hammer blow on the end of the shaft, the first man turned the drill slightly, thus slowly producing a hole. The explosive used was mainly gunpowder, but Dennis introduced dynamite. Later gelignite was also used.

The horizontal workings or levels along the vein were explored by making tunnels on average seven feet high and six feet wide. Where ore bodies were found, the level could be opened out sideways and upwards – called stoping. By careful work, the broken ore was loaded into barrows or tram tubs with the help of gravity. To exploit any ore body extending down below a particular level, a well-managed mine would extend the next level below to meet it. This avoided 'underhand stoping' – getting from below the level – which added the expense of lifting ore upwards, and might also involve the lifting out of collecting groundwater.

Black Tom mine

This part of Snailbeach mine exploited shallower ore deposits, principally worked via the 40 yard

'The Mines, Snailbeach' – a postcard published in Shrewsbury. Looking east along the main siding, Black Tom mine is in the background. As the head frame for the winding gear is turned on its side, this was taken perhaps when the new gear was installed during the Great War. Hewitt's cottage is partly hidden behind trees. The bare track in the background climbs up the dam and turns away at the top.

Horizontal engine at Black Tom mine with Hewitt's cottage in the background. H. F. WHEELLER

level. The ladder shaft to this level is thought to be very old, while the winding shaft is perhaps more recent.[16] Even after Dennis joined the company, Black Tom mine was wound by a horse whim.[17] In 1912 to 1913 the Halvans Co. sank this shaft deeper on an incline below 40 yards, and installed winding equipment and pit work for winding a full and an empty kibble at the same time.[18]

Chapel Shaft (Vessons mine)

This shaft was completed down to the 282 yards level early in 1863, when the brick lining, pit work and winding engine were installed. The winding engine was a marine steam engine, and there were three boilers. According to a valuation, only one boiler was associated with the engine. In 1894 the engine and boilers were all cut out for scrap metal.[19] Although Vessons mine was productive, it seems likely that the shaft fell out of use when the Vessons mine was linked to Snailbeach mine at depth.

Callow Hill mine

No steam engine has ever been mentioned at Callow Hill, and no buildings were ever shown on maps except a pit head frame at the main shaft. The earth works shown on maps suggest that this shaft was wound by a horse whim. When the Snailbeach Mine Co. Ltd and the Snailbeach Lead Mining Co. Ltd operated Callow Hill, all ores were taken to Snailbeach before smelting or sale, and dressing was therefore completed there. Callow Hill mine was drained by an adit which emerged close to Little Minsterley.[20]

THE SNAILBEACH MINERAL VEIN

Vein minerals can occur where there are fault lines and other joints in suitable rocks. Mineral-rich solutions are pushed into the gaps in a series of hydrothermal (in other words hot fluid) movements. Whether or not minerals form in a particular location depends on both the hardness and the chemistry of the host rock – known as the 'country rock' to miners. In some instances mineral formation is related to depth. Various theories have been proposed to explain the process in south west Shropshire: there is clearly much still to be understood. The country rock at Snailbeach is the Mytton beds, a hard grey flag stone.[21]

The following description of the vein is a close transcript from the British Geological Survey.

> The main (Snailbeach) vein runs generally east-west through the Mytton beds. It dips about 60 degrees south, but at the surface near Black Tom shaft it is almost vertical, and in the shaft it runs vertical for about 40 yards. The average thickness is about 8ft but with swells (sometimes almost 40ft and in lower workings often 22ft) and pinches. It is underlain by a smaller or 'back' vein as can be seen just north of Black Tom shaft, this being 3 to 4ft thick. It sometimes intersects the main vein. The back vein is believed to enter the main vein at

> a depth between 40 and 80 yards, where the main vein flattens.
> The bulk of the mineralisation is barytes, but with strings and pockets of witherite and calcite, along with galena, blende, and traces of copper pyrites and iron pyrites. The witherite is not common, but calcite is. As 'spar' it was recovered from the spoil heaps in large quantities for use as building material. It was mainly white in colour, but could also be found in shades of pale amethyst and wine-yellow.
> The galena and blende were also found as pockets and strings in the barytes, but at greater depths. There were virtually none found near the surface, but with depth the galena became more common. At still greater depth the galena was partly replaced by blende, which was said to be most frequent below a depth of 400yards.

The barium minerals were found in best white and dark grades, the latter being stained by mineral oils found naturally in the vein. The dark variety was found to predominate higher in the vein, but despite the staining was still more than 99% barium sulphate.[22]

SMELTING

The Snailbeach smelt works was completed by James Ray Eddy and opened on 11 July 1862 to replace the old works at Pontesford.[23] The old works was built on a colliery, which was at the end of its economic life when Eddy & Son began managing the Snailbeach company's works. The opening of the Minsterley branch railway in 1861 allowed coal from north Wales to be bought and transported to Pontesford and later Snailbeach at a lower price than raising it at Pontesford. It was also of a better quality for smelting, thus helping to increase the output of lead per ton of ore and of coal. Major advantages of moving the smelt works to Snailbeach were firstly that a longer flue to a chimney at a higher location could be built, providing a better draught for the furnaces and improving efficiency; and secondly that the lead ore was carried a shorter distance before smelting, again reducing transport costs (only about three quarters of the concentrates were recovered as metallic lead). All of these factors were extremely important when British lead mines were finding it increasingly difficult to compete against foreign imports. Regular smelting was suspended in 1892 for economic reasons and never restarted (see Chapter Six), although some lead from beneath four of the furnaces was recovered for re-smelting in 1897.[24] After the suspension lead ore was sold to other smelters in Britain, such as Walkers, Parker & Co. Such companies were sent samples of each parcel offered for sale, allowing assay, and the parcels were sold to the highest bidder.

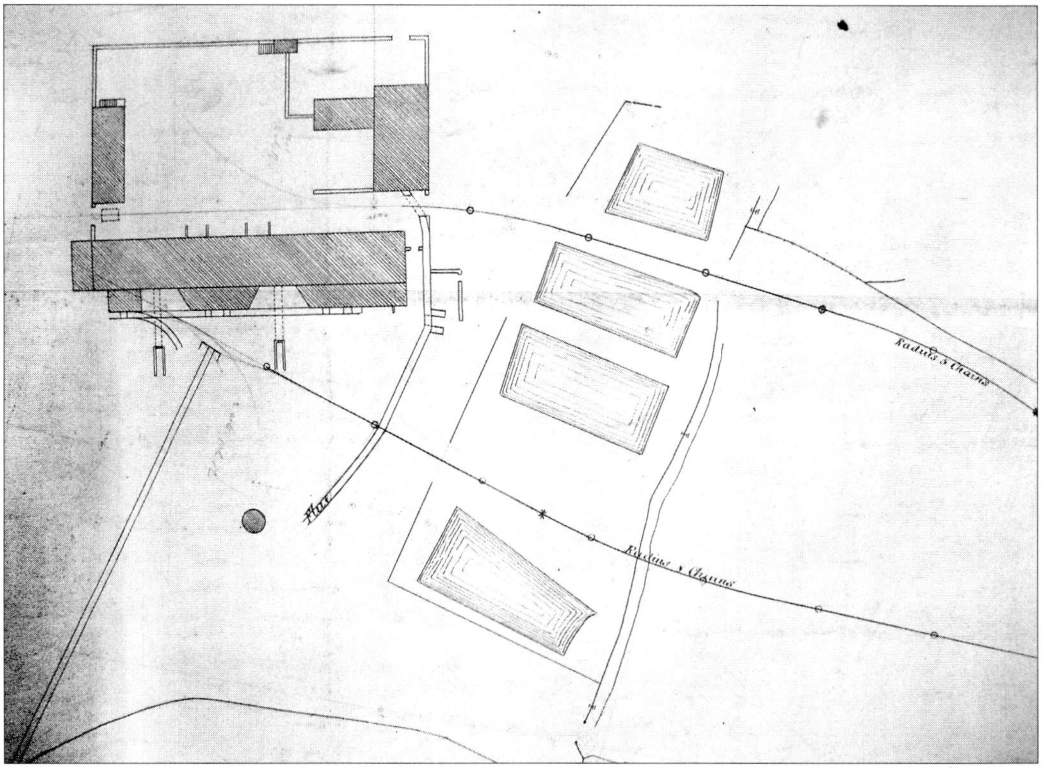

Working plan for sidings at the Snailbeach smelt works. SHROPSHIRE ARCHIVES 4743/48

The slag house viewed from inside the smelt works yard, 2008.
K. LAKE – IA RECORDINGS

Little detail is known about the new works compared with Pontesford, which has had much written about it, but there can be no doubt that the Snailbeach works was equipped and operated in much the same way. It had five reverberatory furnaces inside a long hall, with the flues emerging on the east side of the building. The ore was brought to the works on this side too.[25] At Pontesford, the Eddys replaced an old bellows-blown slag hearth with a Castilian furnace blown by a steam engine. Slag recovered from the main furnaces contained valuable lead, which could be recovered economically by re-smelting in this secondary process. The furnace and engine were probably moved to the new works – certainly a steam engine was in use there. The building in the north-west corner of the new works housed the slag furnace, and a flue leaving it is shown on Dennis's railway construction plan. It has been suggested that the building is so large that there could have been several such furnaces. But there seems to be no justification for this when only one was needed at Pontesford. The smaller annex to this building, projecting to the south, possibly housed the blowing engine.[26]

The lead pigs were stacked in the works yard, ready for hand-loading onto railway wagons (previously carts) for delivery.[27] Coal brought from Pontesbury was probably unloaded direct from the railway wagons at the furnace building, ready for shovelling into the grate at the front of each furnace. Coke was required for a Castilian furnace.

The flues were built from brick on the cut-and-cover principle. They were about six feet high inside, and could be entered by doors set at intervals. The long flue of the Snailbeach works gave improved potential for condensing the metal-laden vapour as 'flue dust', which was collected and re-smelted for sale. Smelting was stopped occasionally, and a company of men would enter the flue to scrape the condensate from the brickwork. According to a report by Henry Dennis, water from the reservoir was released to flush the dust down the flue and into the settling pits arrayed on the north flank of the works. Unfortunately no clues have been found to show how this worked. The furnace flues have been located on the site, but nothing to indicate how the water reached the settling pits from the main flue. The circular brick structure which joins the furnace flues to the main flue has been said to be a 'beehive condenser', but might simply have been a manifold. If it was a condenser, it is not known how it would have worked. There was a condensing house on the north side of the mine, which is identified on a plan dated 1872.

The small building on the south side of the works is known locally as the Smelt House. When the works was operating, the ground floor contained offices, and at its east end a weighing machine. There was an outside staircase to the upper floor, which was a rest room for the smelt workers, known as puddlers, who worked under a foreman smelter.[28] When smelting stopped, the building was rented as a dwelling. Later it was privately owned and occupied. It is now named Lower House, and in recent decades has been altered and extended. The smelt works site (excluding the house) is a scheduled ancient monument.

The smelting process practised by the Eddys at Pontesford was probably continued with little if any difference at Snailbeach. It is described in great detail by Moissenet, as are the furnaces, tools, equipment and much more. Each reverberatory furnace was worked by a team of four, in alternating pairs, usually in twelve hour shifts running end-to-end. They worked two hours on and two off. The furnace was carefully designed and built to throw heat onto the charge, but was also provided with a series of small doors, allowing controlled air cooling as well as observation and working with hand tools. Each charge of 24 cwt of concentrates was dropped through a hopper in the top of the furnace and spread evenly on the sole. It was worked in a carefully tended cycle lasting nine hours, with the tapping of the lead beginning after 8½ hours. The lead was poured into moulds suspended from a steelyard, so that each piece (or pig) weighed exactly 140lb. The yield from the 24 cwt charge was 14 pigs. Any minor maintenance of the furnace was then done, and smelting of the next charge begun. The shift pattern ran from 5.00 a.m. on Monday to 11.00 a.m. on Saturday, consisting of nine shifts of twelve hours and one 18 hour shift. This allowed 14 nine-hour smelting operations per furnace per week. Left in the reverberatory at the end of each charge was 3 cwt of slag, containing about 40 percent lead. The coke-fired Castilian furnace was used at intervals, and could treat 100 tons of slag in a week, recovering 32 percent lead. Flue dust was collected as a muddy material, which first had to be dried. It was then sintered in a reverberatory furnace and reduced in the Castilian furnace. At Pontesford, about half of the fume could be recovered as flue dust, and in turn this would yield 48 percent lead.

USES OF THE METALS AND MINERALS

Lead (Pb). At Snailbeach, the majority of the lead ore was lead sulphide, but at times lead carbonate was found. The Burr brothers of Shrewsbury, and George Burr alone when the SDR opened, were the best customers for Snailbeach lead. They bought pig lead, which from Snailbeach was suitable for making red lead (lead oxide), used extensively in glass making, and for red lead paint. In 1820 Thomas Burr had developed a hydraulic lead extrusion press for making lead pipe. He then patented it and later adapted it to extrude lead sheet.[29] This enabled the Burr brothers to thrive as manufacturers to the building trades. They also made lead shot ammunition using a shot tower. The Burrs took a quarter of the Snailbeach production but also bought from other Shropshire mines. Other buyers had special uses for Snailbeach products. While the Burrs declared Snailbeach slag lead useless for anything other than as common lead, J. & H. Lloyd of the Red Lead Works, Handsworth, bought Snailbeach slag lead specially, but no evidence has been found as to exactly what they used it for. Best & Lloyd of Handsworth bought J. & H. Lloyd in the 1880s, and bought Snailbeach flue dust lead (see above) for glass making.[30] In 1882 a company named Adkins said that nothing other than flue dust lead would do for their work.[31] In 1858, Walkers, Parker & Co. of Chester told the Snailbeach company secretary '... *It is quite a mistake if you think Snailbeach lead any better than other lead (except for glass making). The White Grit lead makes good white lead whereas your lead is utterly unfit for that purpose.*'[32] White Grit was another of the south west Shropshire lead mines.

The contradictory findings of Burrs and J. & H. Lloyd, and the comment from Adkins, show that various industrialists had proprietary metallurgical knowledge developed for their own products. But certainly it is the degree of purity of the lead, and the particular impurities that it contains, which makes it suitable or otherwise for making various lead compounds. Leaded glass (flint or crystal glass) is tinted by traces of copper pyrites found in lead oxides used in making it. Snailbeach lead was exceptionally free of this critical substance. About Snailbeach lead, Moissenet commented: '... *its great purity is very good for making sheet lead and red lead.*' Snailbeach best lead was almost 100 percent pure, and was therefore soft.[33] From the late eighteenth century, while traditional uses of lead, such as in pewter, decreased, new and increasing uses emerged. At that time, a demand emerged for soft lead to make chambers used in the lead chamber method of sulphuric acid manufacture. From the mid-nineteenth century lead became a major component in new alloys made for medium speed machine bearings. From late in the same century it was needed in large amounts for use in lead-acid accumulators. The manufacture of lead compounds was well established by the eighteenth century, falling into three broad groups – lead carbonate, lead oxides and lead acetate – having various uses in the paint, glass, pottery and textile industries.

Zinc (Zn) is the twenty-third most abundant element in the Earth's crust. It is usually found as sphalerite, or zinc sulphide. Zinc production across the world is exceeded only by iron, aluminium, and copper. The uses of zinc have changed little if at all since Snailbeach mine was active. About three-quarters of the world's zinc is consumed as metal, mainly as a coating to protect iron and steel from corrosion (galvanized metal), as alloying metal to make bronze and brass, as zinc-based die casting alloy, and as rolled zinc. The remaining quarter is consumed as zinc compounds mainly by the rubber, chemical, paint, and agricultural industries.

In current use, galvanising typically means hot-dip galvanising, a process used to coat steel or iron with zinc. This is done to prevent rusting of the ferrous item. Hot-dip galvanized steel has been used for more than 150 years. The value of hot-dip galvanizing stems from the relative corrosion resistance of zinc, which, under most service conditions, is considerably better than iron and steel. In addition to forming a physical barrier against corrosion, zinc galvanizing cathodically protects exposed steel. It is also favoured because of its low cost, ease of application, and low maintenance.

Barite (barium sulphate $BaSO_4$). This white rock, which is often found discoloured, is an earthy metal compound. It was best known by the name barytes in Shropshire both mineralogically and commercially, but bore other names – heavy spar, brites, britus, and perhaps others. One of its earliest uses was as a component of ceramics, as with witherite (see below). Josiah Wedgwood used it in Jasperware, although no evidence has been found that Wedgwood used the mineral from Snailbeach. There is evidence of either barite or witherite being sought from Snailbeach for ceramics use in the 1830s. By 1822 barite was used in paints, and later in numerous other ways, including in textiles, paper, card and rubber. Not all mines recognised its commercial value until the mid-nineteenth century, and until then the vast majority was dumped on the spoil heaps. Its use grew in the late-nineteenth century, stimulated by lower cost transport. It is opaque to radiation, and in the twentieth century it was adopted as the 'barium meal' in medical radiography. Several tons of it were mined specially from Perkin's level at Snailbeach in 1957, for use as radiation shielding when there was an incident at Windscale nuclear power station. It is used mixed with clay to form drilling mud in the oil and gas exploration industry. With the application of specialist chemistry, the barium compounds can be and have been used in a wide variety of products and industries.[34]

Witherite (barium carbonate $BaCO_3$) is whitish to pale yellow or pink. It is rarer than barytes, and is more valuable because it reacts easily with acids, making it cheaper to process and form various barium compounds. It has been used in ceramics. Other than this it has little use in its natural state. In refined form it is used in pottery glazes and in enamelling iron and steel.

Calcite (calcium carbonate $CaCO_3$) is one of the most common minerals on Earth, forming about four percent by weight of the Earth's crust. It can form rocks of considerable mass, such as limestones and marbles, and is significant in igneous, metamorphic and sedimentary rocks. Calcite forms the major portion of many hydrothermal veins.[35] It is used as a filler in rubber and textile goods, in the ceramics industry for the preparation of glazes, in paint, and in insecticides. It is also used in the manufacture of cement, calcium carbide, metal polish and sometimes also as a fluxing agent. A flux is a substance that lowers the melting temperature of a material. High quality crystalline calcite is used for making prisms and lenses in optical instruments.[36] The construction industry is the primary consumer of calcite in the form of limestone and marble. These rocks have been used as dimension stones (i.e. worked to accurate shape and size) and in mortar for thousands of years. Limestone blocks were the primary construction material used in many of the pyramids of Egypt and Latin America. Today, rough and polished limestone and marble are still used in prestige construction.[37] At Snailbeach, it has been known as spar, calspar or feldspar, and was often used as decorative gravel and pebbledash in the building trade.

Silver (Ag) is of course a well known precious metal, having a brilliant white metallic lustre and a characteristic sheen. In 1882 and 1883 deposits of lead ore were worked at Snailbeach from which 1,469 ounces of silver were extracted.[38] No other silver production is on record for Snailbeach, and this is because the usual amount of silver per ton of lead (below two ounces per ton) was too low for economic extraction. Some mining historians suspect that Shropshire lead mines including Snailbeach were silver-rich close to the surface, and were exploited as such by the Romans. These shallow workings, certainly at Snailbeach, are no longer evident, either having been trenched away or buried by spoil, so it is unlikely that the theory will ever be proved.

There is evidence that mankind had learned to separate silver from lead by 3,000 BC. Silver is found in ores such as argentite and horn silver, but the principal sources are lead, lead-zinc, copper, gold and copper-nickel ores. The metal is either recovered from the ore, or during the electrolytic refining of copper.

Sterling silver contains 92.5 percent silver, the remainder being copper or some other metal, and is used for jewellery and silverware where appearance is important. Recently about 30 percent of silver produced was used in the photographic industry, mostly as silver nitrate. With the rise of digital photography this use is in decline. Silver is used in dental alloys, solder and brazing alloys, electrical contacts and batteries. Silver paints are used for making printed circuits in electronics equipment. Pure silver has the highest electrical and thermal conductivity of all metals, and has the lowest contact resistance. It is used to make mirrors, as it is the best reflector of visible light known, although it does tarnish with time.[39]

Fluorspar (fluorite CaF_2***)*** or calcium fluoride was produced at Snailbeach between 1874 and 1879. It is found in a variety of geological environments. Pure fluorspar is colourless, but a variety of impurities give fluorite a rainbow of different colors, including green, purple, blue, yellow, pink, brown, and black. It has a pronounced cleavage, which means it breaks on flat planes. Fluorite crystals can be well formed, beautiful and highly prized by collectors. Fluorspar can be used as a flux in making steel, glass, enamel, and other products. Modern consumption of fluorspar is mostly for the production of hydrofluoric acid and aluminium fluoride (AlF_3). Hydrofluoric acid is a key ingredient for the production of all organic and non-organic chemicals that contain the element fluorine. It is also used in the manufacture of uranium. AlF_3 is used in the production of aluminium.[40]

Inside the smelt works flue with Shropshire Caving & Mining Club member Steve Holding in 2008.
K. LAKE – IA RECORDINGS

Snailbeach mine in 2004 – the wagon loading bay behind the locomotive shed. I. COOPER

CHAPTER EIGHTEEN

Postscript

This closing chapter is intended to complete the picture and bring readers up to the present day (2017). It deals firstly with Shropshire County Council's disposal of their old railway land and the Callow Hill quarry; and secondly with the new interest in the railway and the important mining remains at Snailbeach, which has grown in recent decades. Drawing an interest in the mine, the quarries and the railway together can lead to different and even surprising perspectives on these subjects.

By 1986, the SCC owned Callow Hill quarry and the lower railway track bed outright. This was now an accessible, self-contained quarry with possible commercial interest, but one improvement to lorry access was yet to be made. In June 1986, the SCC bought a strip of farm land about five metres wide beside the railway embankment directly south of the A488. This allowed the quarry access road to bypass the entire embankment.[1] Sale of the property followed, and by 1988 the freehold of Callow Hill quarry was owned by Tarmac Western.[2] Under a deed of January 1999, the entrance to the quarry access road from the A488 was improved further, and the bridge over the old railway track bed serving Woodhouse Farm was replaced by a new one on its north-east side. In recent years the freehold of the quarry was listed as owned by Tarmac Roadstone Ltd and then by Tarmac Quarry Products. Now including the old railway track bed, it was owned in 2009 by Tarmac Ltd of Wolverhampton. The railway property north of the A488, mainly used by the SCC as a storage depot for many years, was bought in 2006 by two residents of Pontesbury. By 2009 it was named the Wharf Business Park.

In the 1980s, there was growing concern about the future of Snailbeach, especially about how the historically important array of mining buildings, works and artifacts might be conserved for the benefit of future generations. None were more concerned than those residents whose ancestral families had worked Snailbeach mine and other mines nearby. Various civil engineering concerns, notably Wardell Armstrong, were hired by the SCC to make surveys and produce reports. There was interest in the railway too, but from outside the county. An attempt had already been made in 1977 to lease part of Snailbeach

Snailbeach mine, 14 February 2009. The old crushing engine house and boiler house complex to the right (note the stump of the disused chimney), with the compressor house and central boiler house chimney in the left background. The photographer is standing in the carpenter's yard.
S. A. LELEUX

In 2009, a Snailbeach District Railways (former Granham's Moor Quarry Co.) hopper wagon was restored, using an old, rotten underframe donated by the Talyllyn Railway. Mainly using old photographs, a very creditable rebuild was completed by Barry Ellis, pictured here with the wagon. N. SOUTHWICK

mine as a centre for narrow gauge railway locomotive preservation, but nothing came of it.[3] Eight years later, the *Shrewsbury Chronicle* garnered the following report from the railway enthusiast journal *Branch Line News* of October 1985: *'In connection with a study by Shropshire County Council of the recreational development of the Stiperstones Hill mining area, a "voluntary marketing manager and assistant civil engineer with an existing narrow gauge rail group in Wales" has put forward the possibility of restoring the 2ft 4in gauge Snailbeach District Railways, closed in 1959. The loco shed at Snailbeach could be easily restored but he admits that there are many problems in the way between there and Pontesbury.'* Considering how quickly and easily the track bed had changed hands shortly before – and that one third of it still served the quarry – this was rather an understatement. In due course, the SCC held public meetings at Snailbeach to discuss the conservation plans for the mine and the village. The SCC's notice of a meeting held in July 1990 summarised Wardell Armstrong's recommendations following its Phase 2/3 Report. No mention was made of restoring the railway – scarcely a surprise – although conversion of the locomotive shed to a workshop or study centre was suggested.[4]

Again in 1990, a Chartered Civil Engineer Graham Fairhurst and some friends also investigated rebuilding some of the SDR, but as a two feet gauge railway using equipment they already had. Tentative options included either a line from the A488 at Pontesbury to Callow Hill or around the Snailbeach mine area, on which public rides would be available. Meetings with the Snailbeach mine project officers were held concerning the second option, but it was very clear that visitor access would be difficult, and operation would be restricted to specific open days. Importantly for the Snailbeach conservation project, the proposers also felt that they would not be able to recreate the SDR as it had been, which they felt that the site deserved, and on considering all these and other points the suggestion was dropped.[5]

Since then, much public money has been spent preserving and consolidating the mining remains, including the locomotive shed, and public interest has increased. In 1996, some members of the Shropshire Caving & Mining Club founded the charitable Shropshire Mines Trust to preserve Tankerville mine. The Trust has since expanded its interests to cover mining all over Shropshire, and now manages Snailbeach Mine. The Trust is important to Snailbeach residents for managing the remains sensitively, and providing controlled access and information to visitors.

Even before the 1990s, it would not have been practical to replace the railway. Parts of the track bed had changed in their use drastically since the line was taken up, and the remaining bridges were out of order. Leaving aside the quarry access road, it is often forgotten that even before the SDRCo liquidated its assets, the sanitary authority took advantage of the vacant strip of land to provide Snailbeach with a main sewer. The County Council continued to use Pontesbury sidings for storage. Then in the 1990s, Doug and Mary Challinor built a house on the northern part of the Snailbeach wharf, between the weighbridges and Hughes's bridge.[6] A new railway could never reach the old terminus at Snailbeach, and this would hardly be interesting. Snailbeach, in any case, could not cope with the stream of visitors that a tourist railway can attract. The Snailbeach line has gone for ever, and it is best remembered for what it was and what it did, and the life and work that went on around it.

Regardless of the railway's passing and the dwindling amount of visible remains, railway enthusiasts continue to be drawn to Pontesbury and Snailbeach, to get a glimpse of where the line was, and the mine and the country it served. The 'railway enthusiast' hobby, whatever it might best be called, has existed for a long time. As far as visitors to the Snailbeach railway are concerned, they were visiting as early as the 1930s, if not earlier, as shown by the few photographs by A. W.

Croughton and others seen in this book. Interest in the SDR seems to be rekindled with successive generations. Before the Second World War, travel and photography were relatively expensive, and such a hobby was limited to the few with sufficient money and time on their hands. The post-war economic boom has made such things widely accessible, leisure time (especially paid holidays) has increased, and railway enthusiasts are relatively numerous. Since the 1950s the numbers of people visiting the line have grown, and many photographs were taken in the tractor haulage years.

The author first visited Snailbeach and the railway remains in 1971 as one of those enthusiasts, and it eventually resulted in this book. A similar increase in interest has happened in mining history, and in this context, Snailbeach mine. The author has since become one of those mining enthusiasts, as a result of his interest in the railway. The mine and the railway were mutually dependent for their success and survival, especially at the height of the development of Snailbeach mine, so it is important for those with an interest to take a practical and historical view of the transport and the mining together. This way of looking at the railway (and this history of the line) reflects the history of early railways, as described in Chapter Two. Along the way, I have briefly linked this local history to national and international events and trends; in mining and the minerals and metals trade; in politics, economics and social change; and the effects of wars on these things. I hope readers have found this approach to the history of the Snailbeach District Railways interesting and informative.

ALONG THE LINE TODAY

As has been outlined, the railway land has been parcelled up and sold, with old railway boundaries being removed and new boundaries put across and around it, making it less identifiable as a railway route. The following description of the recent scene is to help newer visiting enthusiasts (and those merely curious) to identify the remains, but it must be stressed that all of the land is in private hands. The author does not condone trespass, and emphasises that anyone following this account should respect the rights of the current landowners, including their privacy. The fact that any particular plot of land is mentioned and described here does not mean that the landowner allows public access to it for any purpose at any time. Any visitors who would like access should first seek permission.

The Pontesbury exchange sidings and northern terminus of the SDR occupied the property north of the A488. After sale by the SCC into private hands, this became known as the Wharf Business Park, but since then the name appears to have changed. At its north end, the plot is bounded by the old Hanwood Junction to Minsterley branch line, which has for many years been in public use as a footpath. The path can be joined from Station Road, Pontesbury. As late as the 1990s, it was possible to see the western boundary wall of the narrow gauge sidings from this footpath. The stone built retaining walls in the cutting at the east end of the junction, on which a standard gauge siding stood, were also still in place.

South of the A488, the embankment leading towards Callow Hill remains, but the entrance to the new quarry road bypassing the embankment is entirely new. In the cutting approaching Callow Hill quarry, a new accommodation bridge carrying the private road to Woodhouse Farm stands over the old railway track bed. The old bridge stood to the south-west side of the new one. Passing the quarry, there is nothing which even hints that a railway once passed through. Neither is any quarry equipment in place, except for the base plate of a lorry weighbridge. Proceeding roughly south to the Callow Lane, the track is on a shelf cut into the hillside. Plenty of excavated rock face can be seen. Nothing at all remains of the railway smithy. At the Callow Lane, the realigned public road appears to cut somewhat below the level of the railway line

The western perimeter wall at Pontesbury sidings in 1993, viewed across the field bought in 1906 by the SDRCo. At this location the wall is now retaining growing tree roots as well as heavy earthworks and is bulging noticeably.

The entrance to the Snailbeach road vehicle yard in 1993. The rectangle of stones in the foreground marks the position of the road vehicle weighbridge. The rods which connected to the machine in the office are under the road to this day.

where they cross. To do this, most of the bridge has been removed, but part of it survives on the west side where the old approach road turned the corner onto the bridge proper. These remains have been filled in and are overgrown.

For a short distance south of the bridge remains, the railway track bed survives with line side trees and hedgerow on either side, making it fairly easy to see, especially in winter. After this section and to the edge of Snailbeach, the track bed has been absorbed into the neighbouring fields, a good deal of which land was wooded when the railway was operating. Any indication of the main line has been thoroughly ploughed out, but along the way, an old pond used in later years as a locomotive water supply marks the Granham's Moor junction. There are various earthworks and concrete artifacts concerned with the Granham's Moor quarry tramway in and close to Maddox's Coppice. The abandoned Granham's Moor quarry workings include various unidentified concrete foundations. Close to Snailbeach, the embankments carrying the branch line to the smelt works have been totally removed, but in the past few years the smelt works has had much vegetation removed as a part of the conservation effort, and much of interest can be seen including track bed on the east side of the furnaces. The works is listed as an ancient monument.[7]

In Snailbeach, much of the track bed is visible, with some parts even having long use as rear access to houses. Alongside the former main spoil heap at the mine, which has public access, the railway route and the Halvans bridge can be seen at close quarters. From there to Hughes's bridge, two parts of the line, absorbed into the village hall and Methodist church grounds, are accessible. The railway wharf from Hughes's bridge to the weighbridge is now occupied by a house and gardens known as 'The Sidings'. The remainder of the wharf, which includes the road vehicle yard, has stayed much as it was since the last track was lifted. Much of it can be seen from the road, and the huge retaining wall alongside the road to Crowsnest is particularly impressive. The foundations of the lime shed and wagon shelter still exist. In line with the yard gate, a rectangle of stones marks the position of the cart weighbridge. Opposite this, across the road, the weighbridge office and a small piece of land are now part of the property named 'Resting Hill'. The office has been redeveloped a good deal for new uses but is still recognisable. Inside, an original iron fireplace has been preserved.

The branch line from the wharf to Snailbeach mine is private property but can be walked by the public. This plot ends at the spoil tip bridge near the locomotive shed, and the mine property, managed by Shropshire Mines Trust Limited, begins at the same point. There is track and points with counterweighted levers outside the locomotive shed. Little else remains of the sidings on the mine, but some track and track bed can be found. In the 1990s, metalwork was found embedded in the ground at the head shunt close to the magazine. This was a remnant of the point leading to the coal yard. At the end of the head shunt is a walled area, which was used for tipping minerals into railway wagons. The raised coal yard next to the 1881 boiler house can be closely

The Snailbeach weighbridge house in 2014. It has been much extended and the old goods doors replaced with French windows.
MICHAEL MESSENGER

viewed from ground level. Parts of the incline serving the stationary engines on Resting Hill can be seen either from the road close to the bottom or from the platform at the top, although a house has been built across it part way up. It is relatively easy to see the mining remains at Snailbeach, now that important derelict buildings have been stabilized and the vegetation is under control, and there is plenty of public information to guide visitors. Unfortunately, through this 'management', much of the atmosphere has been lost, the main spoil heap has been removed, and various buildings, for example the locomotive shed, do not look as they should. To see things as they were, photographs must be relied on. Nevertheless, the mine is worth a visit to get a first hand idea of the mine and the railway working together.

I hope that visitors will gain extra interest in these remains through having read this book. Please go carefully and safely, and show consideration and respect for the residents of Snailbeach and others along the route, several of whom I count as good friends.

Adhesive weight
Weight of locomotive (without fuel and water) bearing on the driving wheels and creating adhesion (grip) and traction.

Axle loading
Maximum weight carried on a single pair of wheels.

Buddle
Wooden frame or other device (rectangular or circular) over which a flow of water passes to separate ore from waste.

Bridge rail (of railway)
A rail of rolled iron or steel having an inverted channel section, with a flange either side at the bottom for fastening to a sleeper.

Check rail (of railway)
Short piece of rail (not carrying locomotives or rolling stock) preventing wheels being misdirected at a point or crossing.

Coupling rod
Bar or rod linking locomotive driving wheels to ensure they rotate at the same speed and can be driven together.

Crossing (of rails)
A special construction in railway track, allowing wheels to be carried smoothly across rails running in a different direction.

Crossover
Short length of railway track joining two parallel tracks.

Deposited Plan
Plan for public scrutiny of railway scheme to be approved by Act of Parliament.

Dressing floor (of metal mine)
An area where minerals are cleaned and separated, often using machines, for further processing or sale.

Dog spike (of railway)
Special large malleable nail, used for fastening e.g. bridge rail, flat bottomed rail to wooden plugs or sleepers.

Edge railway
A railway where the wheels are flanged to retain them on rails held vertically, or on edge. Modern railways are an example.

Gauge (of railway)
For the modern flanged-wheel system, the distance between the two inside faces of the rail heads.

Head shunt
Railway track terminating several sidings which allows rolling stock to be transferred between them.

Level (of mine)
A horizontal (or near-horizontal, to promote drainage) working or tunnel, usually for exploration and winning of minerals.

Mean effective pressure (of steam locomotive)
Value between 75% and 85% of boiler working pressure, allowing for pressure drop between regulator and cylinders, and boiler not being maintained at maximum pressure.

Narrow gauge (of railway)
Less than the prevailing standard gauge (see below).

Passing loop
Siding joined by a point and crossover at each end to a main line, by which trains may pass each other or a locomotive may pass from one end of its train to the other.

Pinch-bar
An iron bar with one end shaped by pinching into a blunt wedge. This end positioned between railway vehicle wheel and rail makes the bar into an efficient lever. By this means, the wheel can be turned and the vehicle moved with little effort

Plankway
An early tracked transport system in mines, where two rows of wooden boards were laid parallel. The gap between them guided wheeled mine tubs which had a guide pin fitted.

Plateway
An early tracked transport system having flat running surfaces and guide flanges for wagons having plain wheels.

Raff wheel
Wheel, associated with roller crushing mill, which returns partly crushed ore to the top of the mill for re-crushing.

Ruling gradient
Steepest gradient of significant length, limiting the maximum load which can be hauled on a railway.

Screening plant/screens
Machinery using a set of wire mesh screens of varying hole size to grade crushed ore or stone.

Sett
Defined area let by landlord for a mine.

Standard gauge
4 feet 8½ inches in Britain.

Tractive effort
Force exerted in turning locomotive driving wheels – a measure of haulage capacity.

Tyre
Outer ring of railway wheel including running surface and flange.

Weight in working order
The weight of a locomotive with full complement of fuel and water.

APPENDIX ONE
Locomotives on loan

The facts about loans of locomotives and wagons by the Snailbeach District Railways to the Glyn Valley Tramway

A number of authors (for instance David L. Davies[1]) have told the story that the two SDR locomotives *Belmont* and *Fernhill*, along with wagons, were sent to the Glyn Valley Tramway (GVT) for use during its conversion from horse to steam haulage between 1887 and 1889. There is no question that both locomotives were used by the GVT, but the evidence shows that it was at two different times within this three year period. Because the SDR was trading throughout this period, with increasing traffic, and recorded Locomotive Power costs, it is impossible that both locomotives could have been absent from the line together for any significant length of time – in such a case horse haulage or carting by road would have been needed. It seems that in 1887 *Fernhill* was sent to the GVT first and returned, to be followed by *Belmont* between January 1888 and April 1889. There is no evidence in either SDR or GVT records of a loan or permanent transfer of SDR wagons to the GVT, and the works contractors were in fact allowed to use GVT wagons.

There was certainly a visiting locomotive on the GVT in 1887, because the GVTCo's annual return for 1887 records Locomotive Power costs although it had yet to take delivery of a locomotive of its own.[2] (Locomotive Power costs reportable to the Board of Trade included wages and materials for operating and maintenance as well as fuel.) In August 1887, the GVTCo agreed with the permanent way contractor, Croom Johnson, to provide them with a locomotive and crew for their work on the conversion, although no record has been found to say when the work started. It was also agreed that Croom Johnson could use GVT wagons, rendering any loan of SDR wagons unnecessary.[3] No SDR records for 1887 have yet been found saying directly that a locomotive had been sent to the GVT. Especially, the 1887 pay sheets are missing, which would show loading and unloading of a locomotive, but the January 1888 pay sheet shows that Edward Wardman was already the full time fireman, indicating that he had replaced James Hughes by 1887.[4] Clearly Hughes had gone to work for the GVT as a locomotive driver by 1887. Added to this, a letter of April 1887 between a landowner, who was selling land to the GVTCo for the conversion, and his solicitor, stated that he *"... had been informed that Steam Power was now being used on the tramway."*[5] This indicates that Hughes, along with *Fernhill*, might have gone to the GVT by April 1887. And on 19 January 1888 Henry Dennis agreed an extra payment to Croom Johnson for *"Assisting with locomotive – Making temporary crossings for 'Fernhill' and 'Belmont' etc. etc."* It appears to have been work done at Chirk to transfer the locomotives, one in each direction, between the GVT and the Great Western Railway.

It might seem a puzzle that Henry Dennis would think *Fernhill* suitable for use on the GVT when the line was still un-rebuilt, and having light duty rails. In fact the locomotive was used as far up the valley as Dolywern, where it was deemed too heavy for the road bridge over the river Ceiriog and could not safely negotiate the sharp curve in the track. This (and the extra payment to Croom Johnson mentioned above) is the only written evidence yet found which confirms that *Fernhill* was used on the GVT conversion work – and is the reason *Fernhill* was returned and *Belmont* sent. It was reported to another author, the late J. I. C. Boyd, that *Fernhill* knocked the bay window off the Herber toll house near Castle Mill on the GVT. This is a local myth. The bay window was removed for a price of £1 in 1873 to provide clearance for the horse tramway.

A further piece of circumstantial evidence supports the loan of a locomotive to the GVT in 1887. In a draft of the SDRCo accounts for the first half of 1888, an amount £2 10s 10d was noted both as received from the Glyn Valley Tramway Co. and as being 'old', so presumably was related to an earlier accounting period. When compared to other known charges for rail transport of the locomotives, the amount looks realistic for a journey from Pontesbury to Chirk. For instance in 1898 *Belmont* was carried to

Chester for repair at Cudworth & Johnson's works for £6 15s, calculated as 9 tons at 15s per ton.[6] So it is suggested that the money received was reimbursement to the SDRCo for the despatch of a locomotive in about 1887 which was being loaned to the GVTCo. It is hard to imagine what else the GVTCo would be paying the SDRCo for in the normal run of things.

For the second loan, surviving SDR pay sheets relate that an unnamed locomotive was loaded onto the standard gauge railway at Pontesbury in January 1888, and that *Belmont* was returned and unloaded in April 1889. During the absence of *Belmont*, the driver and fireman were paid wages, and there is also a record on the pay sheets of a firebox repair. The pay sheets do not show a similar absence for *Fernhill* in 1888 to 1889, the despatch or return of any wagons, or even the return of *Fernhill* from an earlier despatch.

Correspondence with the Board of Trade in 1888 to 1889 provides detail of the work that Belmont was used for on the GVT. The Llangollen Highway Board wrote on 18 December 1888 to complain that '... *the Glyn Valley Tramway Company are running trains loaded with granite and drawn by a locomotive, which does not consume its own smoke, at the rate of 15 or 16 miles per hour* ...' To comply with tramway operating laws, the maximum speed was eight miles an hour and the locomotives were expected to condense their steam. The first GVT locomotive *Sir Theodore* was delivered on 17 October 1888, which was of legal steam tramway design and had steam condensers. Edward Williams, who at the time was secretary of both the GVTCo and the SDRCo, mentioned in a letter of January 1889 to the Board of Trade that steam power was being used '... *for ballasting and other purposes* ...' The Llangollen Highway Board made a second similar complaint to the Board of Trade shortly before *Belmont* was returned to the SDR in April 1889. The second GVT locomotive *Dennis* was delivered in April 1889, giving the line a working spare and allowing *Belmont* to be released.

Belmont remained working on the SDR while *Fernhill* was gone in 1887, and *vice versa* in 1888 to 1889, because the Board of Trade annual returns for the company show Locomotive Power costs for all three years. Concerning wagons, just as there is no evidence of any temporary removal in the form of loading and unloading records, the annual returns show that there was no permanent loss of wagons from the SDR between 1887 and 1889.[7]

APPENDIX TWO
Engineering details

1. DETAILS OF WAY AND WORKS SPECIFICATIONS

With the exception of quotations, which are given either in italic text or indented paragraphs, these details are a summary of the Specifications document.[1]

Specifications

Rail length 21 feet and approx. 40lb/yard, to be supported at 3ft intervals with an extra sleeper at each joint. Sleepers 4ft 10in x 8in x 4in. Standard gauge sleepers 8ft 6in x 9in x 4½in.

Ballast: stone or furnace cinders to pass through ring 1½in maximum diameter, or other material approved as found in excavation, or waste from mines. Initially to be laid twelve inches thick, track laid on top, then extra ballast added and line lifted to correct alignment.

Excavations and embankments: 14ft wide at formation level double line; 10ft wide single. Cutting slopes 1:1, embankments 1½:1 unless reviewed. Turf or soil recovered and relaid on them.

Fencing: oak posts 6ft 6in x 5in x 3½in at 9ft centres, with larch rails 10ft x 4in x 1¼in and one intermediate post 5ft 6in x 4in x 1¼in.

Field and level crossing gates: oak, charred below ground. *'Chamfered Angels'* [angles] on posts. Joints made with white lead and oak pinned. Ironwork from best bar iron neatly forged and fitted. Gates to shut and fasten themselves. Three coats of oil paint – top coat stone colour. Level crossing gates to carry 'shut this gate' painted on one side of each gate in black letters 3in high. Each gate to have ¾in chain and 3in padlock with two keys.

¼-mile posts: oak, sawn square above and charred below ground. *'The number to be painted and secured to the head of the post.'* Three coats of oil paint. One post per ¼ mile on one side of the railway.

Inclination posts: at each start of change of gradient. Oak, *'... wrought above ground, and charred below ground, the top to be cut in the direction of the inclination of the Railway.'* Three coats of oil paint. *'... the rate of inclination to be marked in black figures three inches in height.'*

Brickwork: Bricks for culvert and skew arches, *'... springers, copings and Plinths or skew Angles of abutments and piers, shall be moulded and burnt of the proper form, to fit the proper angles of the piers, and summering of the Arches.'*

28 February 2012. Wrought iron flat bottomed rail found at field edge close to Granham's Moor junction. Finder Andy Wood holds it up for inspection and photography.

> ... In the construction of the oblique or skew arches ... the courses must be laid in a spiral direction whether they consist wholly of brickwork or partly of brickwork and partly of Stone work. For this purpose the centre must be close lagged to the satisfaction of the Engineer who will cause to be marked thereon the direction of the courses to which the work must be accurately executed. ... The backing shall consist of brickwork laid in Mortar, as before described and shall be built in every respect equal in quality to the outside works and the bricks must be cut so as to make them fit the Arches Accurately. The top of the backing must be sloped each way towards the side walls and if necessary cast iron pipes are to be provided for the drainage.
>
> The price for the brickwork, and stonework is to include all expense of excavating foundations and removal of materials filling in over Arches Coffer dams pumping draining pointing scaffolding centreing temporary piles drains &c.

Mortar to be of fresh burnt lime and clean sharp sand in the proportion 3 sand:1 lime. Concrete to consist of

The road under bridge at Pontesbury in the 1930s – the 'iron work' mentioned in the railway specifications. H. F. G. DALSTON

clean gravel and coarse sand and burnt lime the proportion 5 gravel:1 lime. Tipped from a height never less than 8 feet to minimise levelling, and in maximum 18 inch layers.

Filling in over Arches
The spaces between the spandrill walls and above the Arch, and backing of the bridges, Archways &c shall be filled in with large clean cinders, or screened gravel to the height of the underside of the metalling or ballasting of the roadway which metalling must be of good gravel of the thickness shown on the drawing and the roadway formed as hereinbefore described under approaches to bridges.

Puddle over Arches and backing
After any of the arches shall have been turned and backed to the proper height the Arches and backing must be freed from dirt, dust, loose material, or any other extraneous matter ... must be covered with clay puddle which must be properly mixed up and kneaded before being permanently laid.
After the erection of the spandrill walls they must be carefully coated with clay puddle to the extent directed, and great care must be taken to make perfect the connection between the coating of the Arches and the Spandrills.
It is now known by experience that when the work is well executed the Arches are rendered impervious to water, and the Contractor will be held responsible to make them so.

Stonework
... The face work to be random uncoursed rubble masonry to be laid one header to every two stretchers, the header to be in length not less than twice its depth of face ... All arching of stone, whether for bridges or Culverts, to be formed of stone truly dressed to the radius of the arch.

Ironwork
The plates shall be of the best Staffordshire or Shropshire boiler plate, free from blisters and scales, and all other defects. They are to be of the sizes and dimensions shewn on the drawings or of such other dimensions as may be arranged with or directed by the Engineer for the time being.
The plates to be of regular thickness and carefully flatted or bent to the required forms as shewn in the drawings: the edges shall be planed quite true to the requisite form and dimensions so that the edges of the plates may touch each other truly and over the whole surface and so that the rivet holes may if required, be punched correctly by measurement from the edges only and great care must be taken in riveting on the bull plates that the perfect contract [contact] of these surfaces be maintained.
The rivet holes shall in the first instance be punched smaller that [than] the size the rivets will require and shall afterwards when the plates are put together be rimed [reamed] out to the exact size, and every measure shall be used to ensure that the holes correspond perfectly when the plates are riveted together.
The whole body of the rivets, which are to be rose

headed, shall be heated and sufficient means must be adopted to ensure the rivet being upset in its entire length, so as to fill completely the rivet hole.
The Angle and Tie iron must be of the best quality and particularly it must be of such quality of iron as will not Crack or split with such punching or riveting as may be required, any plate or Angle iron, cracked or split, at whatever stage of the work it may be discovered will be condemned, and must be removed and replaced by other iron of the specified quality.
The plates and all other iron must be so placed in the work that the fibres shall run in the direction of the greatest strain on the iron: the bull plates especially require attention to this.
The rivet iron must be of the very best quality that can be obtained.
The whole of the cast iron is to be of the best description of Cold blast metal solid and free from sand holes, flaws and other imperfections.
All girders will be subjected to such test of their strength as the Engineer for the time being may determine and the means for doing this will be at the expenses [sic] of the contractor.
All the ironwork must be painted with 4 coats of paint of such color [sic] as the Engineer for the time being may choose, the cost of which must be included in the price of iron.

Woodwork
The timber used in this work except where otherwise shewn or described shall be Memel or Baltic, of the best quality, and six months out of the water, seasoned by exposure to the atmosphere, and then brought on the ground, and stacked until such time as it shall be wanted.

The price of laying permanent way to include ballast unloading materials from Railway Wagons at Pontesbury Station, the removal of materials to where required to be used and laying same to Complete Permanent Way.
Level Crossings
All roads crossing the Railway without bridges, are to be metalled up to and between the rails as described for Public roads.
Provide, and fix to each rail, two oak guard timbers 10 feet by 6 inches by 7 inches spiked to the Sleepers with 10 inch wrought iron spikes and secured to rails with 3/4 inch wrought iron distance bolts, four to each rail. The guard timbers to public roads to be the same length as the gates, with spikes and distance bolts accordingly.

2. Details of railway equipment

Agreement of the Snailbeach District Railways Co. with the National Mercantile Investment Co. Ltd dated 15 February 1894 for the sale and purchase of the railway line, lands and rolling stock — First schedule [2]

The following is a verbatim transcription:
Junction (Pontesbury) Goods shed 40' 0" x 25' 4" (outside dimensions), timber built, slate roof, with broad and narrow guage [*sic*] line running either side of internal platform 9' wide
Contents 1 x fixed revolving hand crane on platform, to serve either line, to lift 1 ton. 1 office desk & stool.
Wooden Office. Slated roof, brick foundations, dimensions 9' 4" x 6' 4" outside.
One Iron hand portable second Motion Crane to lift 4 tons (maker W.H. Hughes, Ruabon) Not fixed Iron pillar being broken, also foundation bedding stone for ditto said to weigh 40 tons.
One Turntable (12 feet diameter) and tipping table combined, with gearing complete to tranship contents of broad guage [*sic*] ten ton trucks into Snailbeach District Railway trucks (Maker W.H. Hughes Ruabon)
One spare Cast Iron centre for the above turntable
One Wooden Platform 16' 9" x 6' 6"
Two shovels. Two Iron Crowbars and Two Iron Scrapers
One iron chain 15 feet long (3 inch links) One ditto 20 feet long (5½ inch links) Both these used for shunting purposes
One timber Platform with transverse timber bearings on Massive Stone Wall 84 feet long by 13 feet wide used for discharging contents of Hopper Wagons into the Broad Guage [sic] trucks
One Wooden stage Platform for discharging Coal lime &c from Snailbeach District Railway trucks into road carts N.B. This discharging platform is alongside the main turnpike Road from Minsterley to Salop
Callow Hill. One Blacksmiths shop brick and Stone built, with slated Roof, containing one hearth and fire grate, Outside dimensions 19 feet x 13' 3"
Contents of Ditto Quantity of bolts, timber centres for arching and boring rods
Smelting Works Junction. One iron Water supply Column for feeding Engine with slide valve complete
Top Wharf. One double purchase Iron swing crane complete (excepting the timber Yard Arm) to lift 4 tons, set in massive stone bedding. (Maker W.H.

Hughes's bridge, 2009. The Shropshire Caving & Mining Club excavated the fill under this bridge to check its condition. This view shows that the brick lining was laid spiral fashion. The heavy soot deposit can also be seen. Henry Dennis's bridge building specification was strict, and the inspection found no evidence of leaks, old or new, and there were no loose bricks. Contractor Elias Griffith and his men obviously did an excellent job. Seen left to right are Club members Andy Coyle, Kelvin Lake and Emma Coyle. N. SOUTHWICK

Hughes Ruabon) One Railway weighing machine (by Hodgson and Head [*sic* [3]]) 9' 0" x 4' 0" plate weighing to 7 tons
One Cart weighing machine (by Hodgson and Head [*sic* [3]]) 10' 0" x 6' 0" plate weighing to 6 tons
One Weighing Machine House brick built and slated Outside dimensions 16' 6" x 12' 2"
Contents of same – Fire grate yard arms weights and pillars complete pertaining to the foregoing weighing machines. Also sloping office desk 6' 0" x 2' 0" with two drawers (to lock) and one desk (with lift-up lid) 3' 0" x 2' 0" Office stool and form
One timber built lime shed (on stone foundations) with sliding doors dimensions 10' 6" x 9' 0"
One open shed timber uprights and roof 41' 6" x 9' 9"
Rolling Stock - One Locomotive Tank Engine ("Fernhill") 6 wheels coupled (by Lennox Lange Glasgow) 2 cylinders 12 inch diameter 20 ft [*sic*] stroke. Boiler 3' 3" diameter containing 102 tubes. Water capacity 570 gallons weight 15 tons 7 cwt and when under load about 19 tons. In work
In Stock belonging to the said Engine, 14 new tubes
One Locomotive Tank Engine ("Belmont") four wheels coupled with 2 trailing bogey wheels (by Hughes & Co Loughboro). 2 Cylinders 10 inches diameter one foot three inch stroke – boiler containing 80 tubes Weight 10 tons (estimated) and when under load 13 tons (about) In work
In Stock belonging to said Engine 12 new tubes Two Buffer Lamps, One Shunting Lamp and one Guage [*sic*] lamp
One seated car (to carry 20 persons) 12' 0" x 5' 0"
Thirty-one timber frame Wagons fitted with brakes Hadfields steel wheels and axles to carry 3 tons 8' 8" x 4' 8" x 3' 0" internal dimensions[4]
Eleven Hopper Ore Wagons to carry 3 tons, fitted with brakes, dimensions 5' 6" x 5' 2" x 2' 8" bottom discharging aperture 2' 0" x 1' 0"
Six Wagons 8' 8" x 4' 8" x 1' 4" fitted with brakes to carry 3 tons
Six Timber Wagons 7' 0" x 5' 0"
One Hydraulic jack to lift 10 tons
One Screw jack to lift 4 tons

APPENDIX THREE
Traffic statistics

The traffic tonnages given in the tables below have been independently gathered; in part by Brian Janes from SDRCo annual reports and the minute book; and by the author from the published annual returns of railways, which every railway company incorporated by Act of Parliament was bound to submit by law.

As can be seen with careful inspection, there are some inconsistencies between the two sets of figures, and the majority of them are probably clerical errors made by the accountant, clerk or auditor. Where figures are absolutely consistent, these are in bold. The trends are nevertheless quite clear. Annual reports and minutes do not necessarily give tonnages for all categories, so naturally the total of these (column A) is not necessarily the actual total. Note that for a long time, coal was counted as a mineral.

Tonnages stated in the principal chapters of this book are usually given as round figures to illustrate trends only.

	Tons carried per annum									Tons carried per annum			
Year	Goods	Coal	Stone	Lead ore	Barytes	Zinc ore	Spar	Scrap iron	Total (A)	Goods	Minerals	Coal, coke & patent fuels	Total
1877										78	4724		4802
1878										135	12637		12772
1879										101	12361		12462
1880										133	11690		11823
1881										332	14561		14893
1882										155	15625		15780
1883										223	16354		16577
1884										57	8433		8490
1885										57	4933		4990
1886										2615	1859		4474
1887										347	3720		4067
1888										177	4715		4892
1889										130	5135		5265
1890										266	6562		6828
1891										209	6204		6413
1892										26	4623		4649
1893		2279		1135	12	10	739		4175	85	3989		4074
1894		2837		1897	114	25	715		5588	131	5601		5732
1895		2518		1548	70	93	544		4773	47	4740		4787
1896		2785		1165	30	95	663		4738	94	4844		4938
1897		3376		1143	36	288	892		5735	48	5743		5791
1898		3244		1225	43	235	1005		5752	66	5751		5817
1899		3198		1231	109	94	1148		5780	125	5781		5906
1900										226	5672		5898
1901										206	5776		5982
1902										99	4933		5032
1903										119	4780		4899
1904										78	5265		5343
1905										352	7279		7631
1906										994	21198		22192
1907										798	32897		33695
1908										710	34705		35415
1909										749	37456		38205
1910										678	25113		25791
1911										706	20584		21290
1912			7058	35	308	39	688	127	8255		8873		8873
1913			8857	16	2030		668	3	11574	74	11571		11645
1914	279	725	14086		2435		904		18429	279	18150		**18429**
1915	154	445	13868	7	2715	9	713	16	17927	154	17774		17928
1916	120	518	11395		6935		713	32	19713	120	19493		19613
1917	1115	740	16380		4337		860		**23432**	1115	22317		**23432**
1918	1265	550	15184		3872		894		**21765**	1265	20500		**21765**
1919			15723		2715		473		**18911**	1017	**18911**		19928
1920			8634		1538		462		**10634**	530	**10634**		11164
1921			2066		1218		959		4243	449	4300		4749
1922					1590		1040		**2630**	547	**2630**		3177
1923					1401		1839		**3240**	266	**3240**		3506
1924			199		1879		3255		**5333**	448	**5333**		5781
1925			66		505		4546		5117	255	5139		5394
1926			425				4343		4768	312	5324		5636
1927			2589		372		3841		**6802**	320	**6802**		7122
1928			4821		643		3972		**9436**	111	**9436**		9547
1929			7320		465		3580		**11365**	46	**11365**		11411
1930			7727		59		3316		**11102**	55	**11102**		11157
1931	3	246	7918				2353		10520	4	10504	246	10754
1932	75	140	8035				2586		10836	75	10388	140	10603
1933	31	34	14213				2960		**17238**	31	17173	34	**17238**
1934	47	44	18181				3600		**21872**	47	21781	44	**21872**
1935	148	45	21322				3155		**24670**	148	24477	45	**24670**
1936	198	36	14973				3362		**18569**	198	18335	36	**18569**
1937	137	20	18571				2801		**21529**	137	21372	20	**21529**
1938	109	3	23684				2556		**26352**	109	26240	3	**26352**
1939	84	2	20131				1632		21849				
1940	70	4	17144				46		17264				
1941	31	2	13102						13135				
1942	21	3	12994						13018				
1943	38	2	9877						9917				
1944	16	3	8315						8334				
1945	14	2	8369						8385				
1946	17	2	11778						11797				
1947	16	5	12810						12831				

APPENDIX FOUR

The Buxton Hill railway extension schemes

Two routes were surveyed for an extension of the SDR to Buxton Hill, to the south west of Snailbeach. The schemes have been examined and compared thoroughly by the late Rodney Weaver.[1] As with all earlier extension plans, nothing was actually built. For some time though, a branch of the aerial ropeway built by Shropshire Mines Ltd served Buxton Hill (see Chapter 7). This branch was for carrying stone from a quarry there to Poulton, where it was collected at the road side. There is no other mineral wealth known on Buxton Hill, and the proposed railways can only have been intended to carry stone. The termini of the ropeway and the proposed railways on Buxton Hill were very close.[2] No documentary evidence other than the surveys seems to have survived, and since much about the schemes including their dates can only be speculated, it has been thought best to confine discussion to an appendix rather than the main work.

The two railway surveys were drawn on one plan, of which there are two known copies. One copy has annotations concerning offers for the sale or rental of the land needed for building the railway.[3] The plan is undated, but the annotations can be given an earliest date of 1922. The Joint Lordship of Lloyd and Tankerville previously owned much of the freehold, and held an auction sale of the estate in 1922.[4] This was when the land fell into local hands, as indicated by the annotations. Surveyors with instruments have also been recalled working near Crowsnest Farm in about 1923.[5]

No proof has been found to identify which company (or companies) made enquiries about a line to Buxton Hill, although no company other than Shropshire Mines Ltd is likely to have been interested in building such a railway extension while that company held the mineral rights. The author speculates that the first scheme dates from early in the Great War, when Sir John F. Ramsden would have been prospecting prior to the formation of Shropshire Mines (1916) Ltd. Any such railway scheme was outclassed by Shropshire Mines Ltd's later plan for a ropeway to serve the Bog mine which also had a branch to Buxton Hill, completely eliminating intermediate road transport and transhipment. The main ropeway was operating by 1919. There is no certain knowledge of how long the quarry and branch ropeway were worked. Local opinion is that the stone was of poor quality and consequently the quarry was not worked for very long, but no date can be put on its closure.

The author also speculates that the second scheme was commissioned by H. F. Stephens, on the basis of the datable annotations to the survey mentioned above. Stephens began negotiating the purchase of the SDR in 1920, and took possession in late 1922. Although the ropeway was in place and its Buxton branch perhaps working, Stephens needed traffic for the railway. It is assumed that he was investigating the likely cost, prior to approaching Shropshire Mines Ltd about replacing their ropeway branch serving Buxton Hill. But there were obstacles to such a scheme other than cost. The main Bog to Minsterley ropeway crossed the route of the proposed railway, and the ropeway buckets ran close to the ground. A minimum height for the buckets to travel above fields of ten feet and above roads of 15 feet was specified on a Lloyd lease covering part of the route not far from where the crossing point would have been.[6] This would have left scant clearance for the railway passing underneath. But it need not have been an insurmountable construction problem, since extra ropeway pylons could have been installed either side of the railway, or railway earthworks made. It is likely that continued general operation of the ropeway was a greater obstacle. Shropshire Mines Ltd continued to operate the Bog mine until 16 October 1925, and there is no evidence to suggest that the ropeway (or even the Buxton branch) was closed before that date. In such circumstances, what proposal Stephens might have made to Shropshire Mines Ltd in a bid to secure trade from a quarry of doubtful value is hard to imagine. In 1924 Hayward's stone traffic began on the SDR, so all such expensive schemes could be shelved.

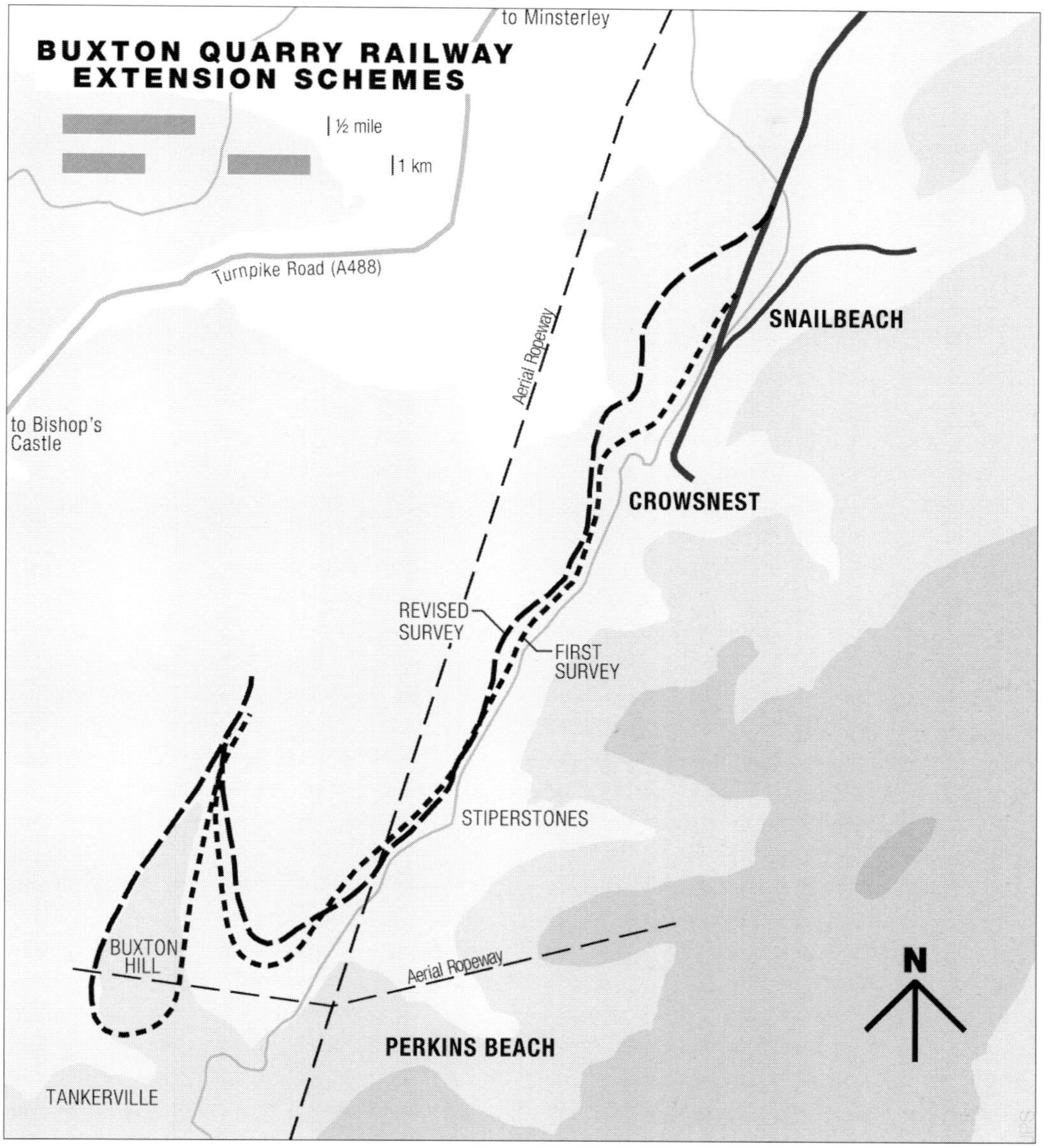
BUXTON QUARRY RAILWAY EXTENSION SCHEMES
½ mile
1 km
to Minsterley
Turnpike Road (A488)
Aerial Ropeway
SNAILBEACH
to Bishop's Castle
CROWSNEST
REVISED SURVEY
FIRST SURVEY
STIPERSTONES
BUXTON HILL
Aerial Ropeway
N
PERKINS BEACH
TANKERVILLE

APPENDIX FIVE

The Boycott Colliery tramway proposal

The earliest known proposal for a railway or tramway to feed the SDR with traffic, other than at Snailbeach mine, was produced by the Asterley and Boycott Colliery company while the SDR was under construction. Boycott Colliery lay roughly north west of Pontesbury Junction, and on the opposite side of the Minsterley branch railway to the SDR. In summer 1876 plans were produced by a Mr J. Freeberne for the new tramway. A few surviving letters indicate both the purpose of the tramway and its importance to plans for developing the colliery and improving its profitability. In the event the tramway was not built.[1]

The Asterley and Boycott Colliery company began operating in early 1874. The company's new colliery manager, a Mr Stokes, had already improved the economic working of and sales from the colliery. He was hoping to find better quality coal measures below the three thin measures usually found in the Pontesbury area. On 26 August 1876 he wrote to the principal investor Viscount Hill's estate manager Alec Hathorn about progress and the new proposals, including the following:

> ...Taking a general view of the whole concern & keeping in mind that there are no less (including our own) than four collieries competing for the trade of this small district, it is necessary that the following works should be carried out to put the Colliery on a firm footing.
>
> To sink at least one shaft on Miss Boycott's land down to the Thin Coal, according to agreement, after which boring should take place to prove the existence of the lower, thicker and better seams. Should these seams prove to exist the future of the Colliery is sure.
>
> The Tramway connecting the Pit with the Snailbeach Line should be laid at once, as an outlet to the Mines when the latter line is opened for trade, & also as an outlet to the General Market in the event of the thicker seams being won. Should this Tramway not be laid I feel hopeless as to the Colliery proving a success. The negociations moreover with the several parties for power to cross & use their land in the construction & working of the Tramway having all but been completed its abandonment will I fear place all parties concerned in an awkward position.
>
> ...I fear to say that the Colliery will show a profit unless the Tramway is laid. The adjoining Colliery has placed a Traction Engine on the road in order to increase his sale so needful did the owner find that something must be done in this direction.[2]

Hathorn supported Stokes's view, and on 29 August 1876 wrote to Viscount Hill:

> I enclose Mr Stokes' letter to me of the 26th inst received yesterday.
>
> It sets forth the true position of matters as they now stand, and as they will be, if the Tramway be not made or even postponed.
>
> ...I have completed all necessary preliminary arrangements with the following persons, vist,
>
> With the London & N Western & Great Western Railway Companies for permission to run under their joint Railway with the Tramway, on payment of £10 per an:
>
> With the Revd Mr Kingsford (one of the Rectors of Pontesbury church) for a portion of his Glebe Land near the junction of the Snailbeach District railway with the above Railway Companies' line to Minsterley
>
> With Dr Eddowes of Pontesbury for leave to pass along his two fields with the Tramway, after passing under the Companies' Railways. The Messrs Salt are acting for both Mr Kingsford & Mr Eddowes, & are preparing the necessary agreements.
>
> ...But apart from what are involved in the above negociations, the making of the Tramway is really necessary, and that it should be done now, so as to be ready for work at the same time as the new Line of Railway to Snailbeach is opened for traffic. If this is not done, other collieries, both near & distant, & now competing for supplying coal to the lead mines, will have established Depots at the Junction of the New Line at Pontesbury, and have made arrangements with both Companies for transhipping their coals from broad to narrow guage, & so competing with Boycott. But if we can forestall these Colliery Companies, & be able to send up our coals without stoppage of any kind, we must beat them out of the supply of the mines.
>
> ...A great deal of difficult & harassing work has been got over, coal has been won, & the Tramway can be supplied with plenty of coal.
>
> Great exertions have been used – are being daily used to make the Colliery a success, but there must be the means provided to make it so – if they are

> properly supplied, I will answer for it that they will be turned to the very best & most profitable account.
> Of course if the deeper seams (the Ruabon seams) of Coal be found below our three known seams, the success of the Colliery will be great & certain. This may be known very soon, if only the required shaft or shafts be sunk in Miss Boycott's land at once. Boring rods are all ready for use... [3]

Hathorn mentions plots of land which would be used for the tramway in crossing the Minsterley branch and joining the SDR. Knowledge of successive ownerships of local land has allowed these plots to be identified, as shown on the accompanying map, although their sequence does not match Hathorn's description literally. Unfortunately the plan produced by Freeberne has not been found, and only a general idea of the route can be given.

A further letter by Hathorn of 8 Sept 1876 to the Viscount adds:

> ...[Stokes] also reports of the Snailbeach Railway "This is again begun & the Manager tells me that there is not the slightest doubt but that it will go on to Tankerville. I pointedly put the question to him & he assured me it would be the case." ... [4]

This indicates that Stokes spoke either to Dennis or at least a senior man at Snailbeach. For some reason, some importance was placed on the completion of the SDR to Tankerville mine (the construction of Railway No.2) by the Boycott colliery management, but in fact this plan had been abandoned by the SDRCo in the spring of 1876 (see Chapter Five), and unless the SDR management was deliberately leading Stokes on, it is a puzzle that he should be told that the Tankerville section was to be built. One answer might be that Stokes was already aware that Snailbeach mine and smelt works was buying its coal from the Denbighshire coal field, which afforded more economical smelting. Boycott Colliery would therefore need access to other mines in the ore field to attract custom. Certainly the SDR would have welcomed the potential of extra traffic. In the event, Viscount Hill decided against all the ambitious development plans for Boycott Colliery, and it is now known that the Ruabon seams do not exist in that area. The tramway proposal was abandoned and the colliery appears to have ceased trading on 16 April 1877, before the SDR opened.

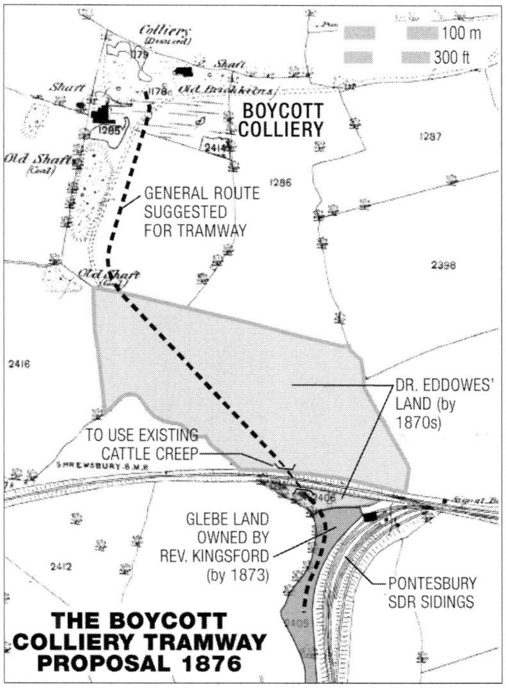

Location and suggested route of Boycott colliery tramway proposal of 1876. The parcels of land mentioned in the correspondence, over which wayleaves had been negotiated, are indicated.

I am especially grateful to Michael Shaw for the information presented in this appendix.

APPENDIX SIX

Contractors' drawings

The drawings presented here are samples of a number of contractors' drawings held in the Longueville Collection at the National Library of Wales. They were drawn by Henry Dennis' engineering office in Ruabon and signed by the contractor Elias Griffith. Other drawings include bridges and culverts, an early plan of Pontesbury Junction sidings, and the station building which was never needed.

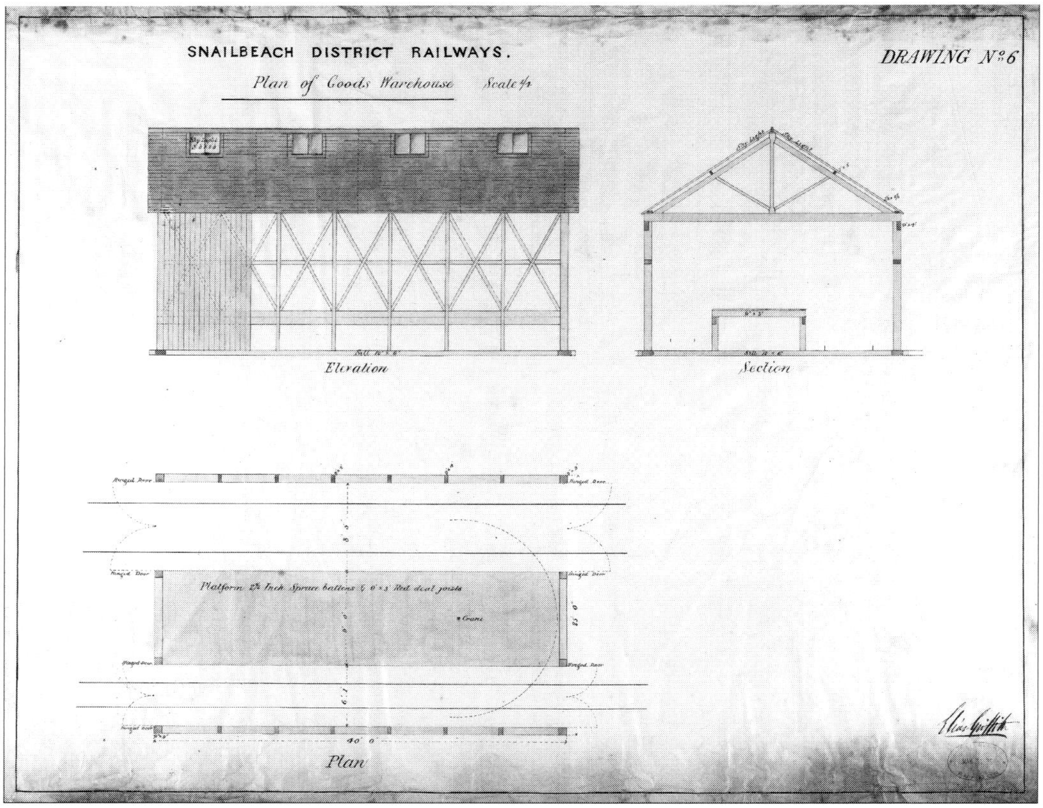

Drawing No.6 is the only image we have of the warehouse, but it is known that it was slate roofed, and the exterior timber gas-tarred a little after it was built.
NATIONAL LIBRARY OF WALES

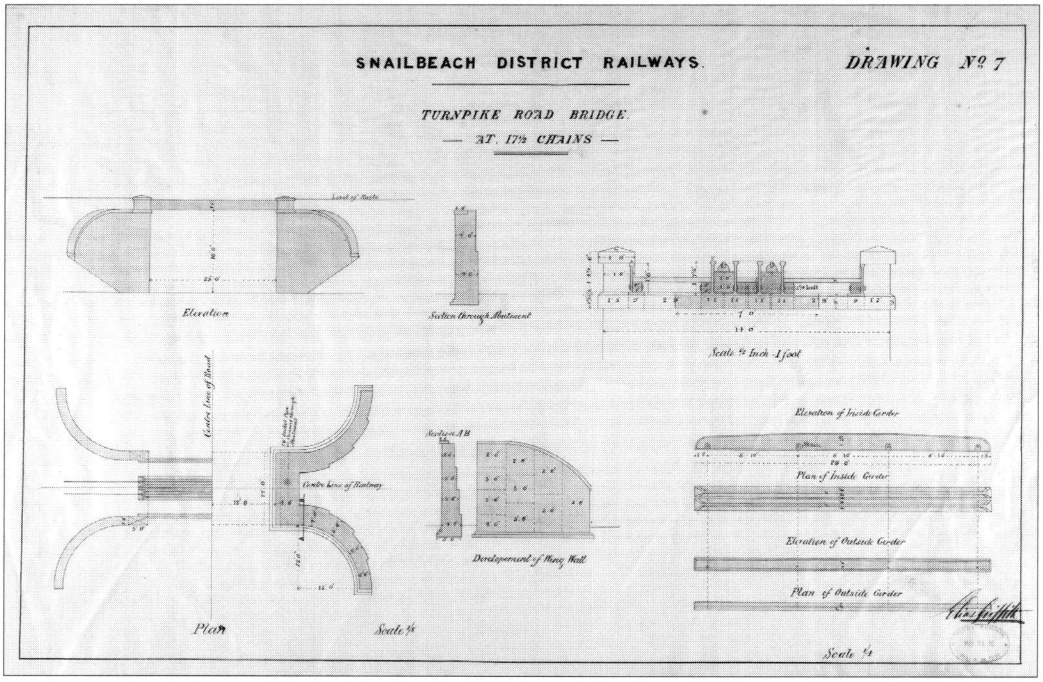

Drawing No.7 of the road under bridge at Pontesbury details the wrought iron bridge work supplied by the Brymbo Iron Co.
NATIONAL LIBRARY OF WALES

BIBLIOGRAPHY

Baker, A. C. & Civil, T. D. A. *Bagnalls of Stafford* (Ashford, Kent: Phyllis Rampton Narrow Gauge Railway Trust, 2008)

Boyd, J. I. C. *The Talyllyn Railway* (Didcot: Wild Swan Publications Ltd, 1988)

Boyd, J. I. C. *Saga by Rail: Great Britain and the Isle of Man* (Usk: Oakwood Press, 2007)

Brook, F. & Allbutt, M. *The Shropshire Lead Mines* (Hartington: Moorland Publishing Co., 1973)

Burt, R. *John Taylor, Mining Entrepreneur and Engineer 1779-1863* (Hartington: Moorland Publishing Co., 1977)

Burt, R. *The British Lead Mining Industry* (Redruth: Dyllansow Truran, 1984)

Burt, R., Waite, P. & Burnley, R. *The Mines of Shropshire & Montgomeryshire* (etc) (University of Exeter Press/Northern Mine Research Society, 1990)

Cox, D. & Krupa, C. *The Kerry Tramway and other timber light railways* (Brighton: Plateway Press, 1992)

Davies, D. L. *The Glyn Valley Tramway* (Oakwood Press, repr.1966)

Francis, P. et al *Never On A Sunday* (Scenesetters/Shropshire Mines Trust, 2000)

Gaydon, A. T. (Ed.) *The Victoria History of the Counties of England, Shropshire, Volume eight* (Oxford: Oxford University Press, 1968)

Gratton, R. & Band, S. R. *The Ashover Light Railway* (Didcot: Wild Swan Publications Ltd, 1989)

Gratton, R. *The Leek & Manifold Valley Light Railway* (Garndolbenmaen: RCL Publications, 2005)

Griffith, E. C. *The Bishop's Castle Railway, Shropshire, 1865-1935* (Author, 1949)

Hughes, S. *The Brecon Forest Tramroads* (Aberystwyth: Royal Commission on Ancient and Historical Monuments in Wales, 1990)

Lewis, M. J. T. *Early Wooden Railways* (London: Routledge & Kegan Paul, 1970)

Milner, W. J. *The Glyn Valley Tramway* (Poole: Oxford Publishing Co, 1984)

Milner, W. J. *Slates from Glyn Ceiriog* (Chester: Ceiriog Press, 2008)

Milner, J. & Williams, B. *Rails to Glyn Ceiriog – The History of the Glyn Valley Tramway* Part 1 (Chester: Ceiriog Press, 2011)

Mitchell, V. & Smith, K. *Branch Line to Shrewsbury* (Midhurst: Middleton Press, 1991)

Morton, A. L. *A People's History of England* (London: Lawrence & Wishart Ltd, 1965)

Popplewell, L. A *Gazetteer of the Railway Contractors and Engineers of Wales and The Borders 1830-1914* (Melledgen Press, 1984)

Redwood, C. *The Weston Clevedon & Portishead Railway* (Weston-super-Mare: Sequoia Publishing, 1981)

Shaw, M. J, et al *Aerial Ropeways of Shropshire* (Shropshire Caving & Mining Club, 2015)

Smyth, W. W. *Metallic Mining* (England: 1877)

Stanier, P. *Quarries of England and Wales* (Truro: Twelveheads Press, 1995)

Strange, P. *The Weston Clevedon & Portishead Railway, a Pictorial Record* (Truro: Twelveheads Press, 1989)

Tonks, E. S. *The Snailbeach District Railways* (Industrial Railway Society, 1974)

Page, W. (Ed.) *The Victoria History of the Counties of England, Shropshire, Volume one* (London: Archibald Constable, 1908)

Wade, E. A. *The Plynlimon and Hafan Tramway* (Truro: Twelveheads Press, 2nd edition, 1997)

Wells, A. G. *Bowaters' Sittingbourne Railway* (NGRS, 1962)

Primary and documentary sources, articles in magazines and journals and books referred to by the author's informants are given as references in the text along with the informant's name.

Abbreviations

CSRMA Colonel Stephens Railway Museum Archive
DHS Transactions of the Denbighshire Historical Society
GVT Glyn Valley Tramway or Glyn Valley Tramway Co.
IRR Industrial Railway Record
NGRS Narrow Gauge Railway Society
NG Revw Narrow Gauge & Industrial Railway Modelling Review
NLW National Library of Wales
NMRS Northern Mine Research Society
NRM PSH National Railway Museum; Selwyn Pearce Higgins collection
OS Ordnance Survey
PA Parliamentary Archives (formerly House of Lords Record Office)
pers comm personal communication
SA Shropshire Archives
S&MLR Shropshire & Montgomeryshire Light Railway
SCMC Shropshire Caving & Mining Club (formerly Shropshire Mining Club)
SDR; SDRCo Snailbeach District Railway or Snailbeach District Railways Co.
SMCo Ltd Snailbeach Mine Co. Ltd
SLMCo Ltd Snailbeach Lead Mining Co. Ltd
TNA The National Archives

Chapter One

1. Victoria County History of Shropshire (1908) Vol I, p.262

Chapter Two

1. Lewis, 1970; Hughes, 1990
2. Boyes, G, ed. *Early Railways 4*, pp.70-84 (Sudbury, 2010: Six Martlets Publishing)
3. Courtesy R. C. Link; Curr, J. *The Coal Viewer and Engine Builder's Practical Companion* (1797); M. J. T. Lewis, pers comm
4. Boyes, G, ed. *Early Railways 4*, pp.177-191 (Sudbury, 2010: Six Martlets Publishing)
5. R. C. Link, pers comm
6. Fairbairn, R. A. *Lead Mine Waggons* British Mining No.54 (Keighley: NMRS, 1995)
7. SA MI 6750 Diary of Thomas Francis Poole
8. Hannaford, H. R. & Price, G. *A Watching Brief at Snailbeach Mine*, Shropshire Report No.58, plate 22 (Shrewsbury, 1995: Shropshire County Council); Shropshire Mines Trust newsletter 71, pp.4-6
9. Lewis, M. J. T, ed. *Early Railways 2*, pp.102-117 (London, 2003: Newcomen Society)
10. Shropshire Mines Trust newsletter 71, pp.4-6
11. M. J. Shaw, pers comm
12. I. Cooper, pers comm
13. Shropshire Mines Trust newsletter 72, pp.2-4
14. D. Poyner, pers comm
15. P. Teather, pers comm
16. SA 2194/4 Engineer's Notebook; K. Lake, pers comm.
17. Author's site visit notes, 1971
18. M. C. Gill, pers comm
19. SA 5982 6/15; PA Snailbeach District Railways Act 1873
20. OS County Series Shropshire sheet XLVIII.6, 1881
21. SA 800/25D Manager's Reports
22. SA 5982 14/15 J. R. Eddy's agreement 1/3/1864
23. SA 5982/34 Lease 24/6/1863
24. Burt, 1984; SA 5982 4/15 6s per ton in 1856; Burt *et al*, 1990
25. SA 5982 14/15 Dennis's agreement
26. Milner, 1984
27. Victoria County History - Turnpiked 1758 Shrewsbury to Minsterley; 1834-1838 Minsterley to Bishop's Castle
28. Burt, 1977
29. Smyth, 1877

Chapter Three

1. SA 5982 14/15 Dennis's contract
2. Courtesy A. C. Baker; obituary in *Proceedings*, Institution of Civil Engineers, London
3. Milner, 1984; Stanier, 1995
4. Courtesy Denbighshire Record Office; OS County Series Denbighshire 25in sheets XXXIV.2,6,10 & 11, 1st/2nd editions
5. Edwards, I. "Slate Quarries in the Llangollen District" Trans DHS 324 (1985), pp.91-120; Courtesy RCAHM Wales; *Cadw* Welsh National Monuments Record 19700
6. Courtesy W. J. Milner; Denbighshire Records Office (Ruthin) QSD/DR/176-177 Deposited plan and section
7. PA; Wrexham District Tramways Act, 1873
8. Courtesy R. C. Link; Klapper, C. *The Golden Age of Tramways* (London: Routledge, Keegan & Paul, 1961)
9. Lerry, G. G. "Henry Dennis: A Cornish Captain of Welsh Industry" Trans DHS 1 (1952), pp.33-45
10. SA 800/43A
11. PA; Snailbeach District Railways Act, 1873
12. R. C. Link, pers comm
13. SA 2194/4 Engineer's Notebook

14. SA 4743/48 Snailbeach District Railways working plan
15. Baker, A. C. "Bagnall and the WDLR 2ft 0in Gauge Locomotives" IRR No.130 (Sept 1992)
16. Tonks, 1974
17. Cuckson, A. S. "BELMONT – the Snailbeach Hughes locomotive" IRR No.171 (Dec 2002); Milner & Williams, 2011

CHAPTER FOUR

1. Griffith, 1949; TNA RAIL 1075/67; Brook & Allbutt, 1973
2. TNA RAIL 1007; Popplewell, 1984
3. Lock, K. C. "The Minsterley Branch Railway" *Shropshire Mining Club Newsletter* No.97 (Mar 1973)
4. Hordley, J. *Turning all our Lead to Gold – An Account of the Burrs* (etc) (University of Birmingham dissertation, 2000); SA 800/25C
5. SA DP396
6. SA DP440; SA DP519/2-3; Bishop's Castle Railway Society web site
7. Burt et al, 1990
8. SA 800/25D
9. SA 1709/6/199
10. SA 5982 14/15 Dennis's contract
11. SA 800; SA 5982
12. Courtesy W. J. Milner; Pritchard, T. W. *Remembering Ruabon* (Bridge Books, 2000)
13. SA 5982 13/15
14. TNA BT 31 611/2567; TNA BT 31 1497
15. TNA BT 31 1618/5490
16. SA 5982/3
17. TNA RAIL 1014/5/17
18. Morton, 1965
19. SA 5982 2/15 SMCo Ltd Minute Book
20. SA DP469
21. SA 800/43C
22. PA; Snailbeach District Railways Act, 1873
23. PA; South Shropshire Railways Bill
24. SA 5982 1/4
25. SA 800/43D
26. PA; Snailbeach Mine Estate Act, 1867
27. SA 800/43B Snailbeach District Railways Prospectus
28. Burt, 1984; SA 5982 4/15; TNA RAIL 1053; SA 5982 5/15

CHAPTER FIVE

1. SA 5982 13/15; CSRMA, SDR Minute Book; PA, Snailbeach Mine Estate Act 1867
2. SA 800/43D; SA 800/43C
3. Courtesy I. J. Brown, *The Advertizer* 6 Feb 1903; Bishop's Castle Railway Society web site
4. SA 800/43B
5. SA 800/43A
6. W. J. Milner, pers comm
7. Milner, 1984
8. Courtesy M. J. Shaw; TNA BT 31/1522/4793; BT 31/2709/14603
9. SA HB5/D/1/1
10. PA; Snailbeach District Railways Act 1873
11. SA 5982 2/15 SMCo Ltd Minute Book
12. TNA BT 285/506
13. SA 800/25D
14. SA 5982 3/4
15. TNA MT 6 180/6

CHAPTER SIX

1. SA 800/43B
2. Courtesy M. J. Messenger; PA Regulation of Railways Act 1842 and later modifying Acts
3. *Shrewsbury Chronicle* 31-8-1877
4. SA 800/43D
5. CSRMA; SDR Minute Book
6. SA 800/43C
7. SA 800/43A
8. TNA RAIL 1053
9. SA 800/42C
10. SA 5982 2/15 SMCo Ltd Minute Book
11. SA 5982 4/4; SA 5982 2/4
12. Cuckson, A. S. "Snailbeach: Railway Staff, the repairs to DENNIS etc." IRR No.180 pp.89-98 (Mar 2005)
13. SA 5982 2/15 SLMCo Ltd Minute Book; TNA BT31/14804/21384
14. PA
15. Courtesy S. Whitehouse; J. G. Vincent Collection
16. Cuckson, A. S. "The Snailbeach Barytes Mill" *SCMC Journal* No.7 (2001) Burt et al, 1990
17. Burt et al, 1990; PA, Snailbeach District Railways Act 1873
18. SA 800/26C
19. SA 5982 10/15
20. PA, Shropshire Minerals (Light) Railway Act 1891
21. TNA Index of Registered Companies
22. Brook & Allbutt, 1973; Willies, L. "A Note on the Price of Lead, 1730 – 1900" *Bulletin of the Peak District Mines Historical Society* (1969) Vol.4 Pt.2 pp.179-191
23. Courtesy M. J. Shaw; SA HB5D/1/3
24. Courtesy M. J. Shaw; SA 3/1A/1/2
25. PA; Light Railways Act 1896
26. SA 484/1592
27. SA DP521
28. TNA MT 14/1
29. Courtesy M. J. Shaw; SA 1033/50-53

Chapter Seven

1. SA 5982 2/15 SLMCo Ltd Minute Book; TNA RAIL 1053
2. CSRMA; SDR Minute Book
3. SA 4155
4. Cuckson, A. S. "The Snailbeach Barytes Mill" SCMC Journal No.7 (2001)
5. NLW; Longueville Collection, SDR Letter Books
6. TNA BT 31 17523/85429
7. SA SC3/1a 1/4
8. SA SC3/1a 1/5
9. Courtesy J. Lawson
10. Courtesy W. J. Milner
11. SA 4743/50; SA DA42/154/2/23; K. C. Lock, pers comm; OS National Grid 1:2,500 1st Edition 1973 SJ 3803-3903
12. Davies, 1966; Milner, 1984
13. A. C. Baker, pers comm
14. SA 4044/37
15. Baker & Civil, 2008
16. SA 5982 2/15
17. SA 5982 1/15
18. SA 5982 8/15
19. TNA BT 31 14691/15895
20. Courtesy B. Janes; Traffic statistics table
21. Courtesy M. J. Shaw; SA DA42/154/2/23
22. Courtesy Industrial Locomotive Society
23. SA SC3/1a 1/7
24. TNA BT 31/23508/145661
25. Courtesy M. J. Shaw; Evans, G. *A Voice from the Hills* (private publ. M. Ridge, c.1993); Ramsden, J. *The Employees, The Shropshire Mines Ltd* (Company publ., Dec.1919) SCMC library KN6
26. TNA BT 285/506
27. Cox & Krupa, 1992
28. M. Rowson, pers comm
29. E. Rowson, pers comm
30. PA; Ministry of Transport Act 1919
31. TNA Parliamentary papers 126.243
32. Mitchell & Smith, 1991; S&MLR freight table
33. P. Shaw, pers comm
34. Courtesy B. Janes; TNA MT 6
35. M. Chidley, pers comm
36. CSRMA
37. T. Burnham, pers comm

Chapter Eight

1. CSRMA; M. J. L. Iggulden, pers comm
2. CSRMA, SDR Minute Book; TNA BT 285/506
3. SA 4155 Halvans Co. Ltd Day Book
4. Courtesy SA, Kelly's Directories of Shropshire; E. Rowson, pers comm
5. M. Rowson, pers comm
6. Courtesy I. J. Brown; newspaper obit., publ. unknown
7. M. Challinor, pers comm
8. R. G. Cooke, pers comm
9. D. Challinor, pers comm
10. SA 4155
11. Courtesy B. Janes; Traffic statistics table
12. SA 6000/19510; SA 2478/9-15
13. TNA Index of Registered Companies
14. SA 2478/47; SA 2478/50-77
15. SA 2478/72
16. TNA RAIL 1053
17. J. Lawson, pers comm
18. SA SC3/1a 1/15
19. Courtesy SA; Wilding's Directories of Shrewsbury
20. G. France, pers comm
21. OS SJ 30 NE (1:10,560) Provisional Edition 1954 *'Whole sheet revised pre-1930'*
22. Courtesy I. J. Brown, *The Quarry Managers' Journal* (1928-9); Terex Pegson Ltd
23. SA SC 3/1a 1/12
24. Courtesy I. J. Brown; Annual Official Lists of Quarries
25. SA 1950/50
26. Tonks, 1974
27. SA SC 3/1A 1/14
28. NLW; Longueville Collection, SDR Letter Books
29. Courtesy B. Janes; NRM PSH
30. Boyd, 2007
31. Weaver, R. "More about Snailbeach" IRR No.70 (April 1977)

Chapter Nine

1. SA SC 3/1a 1/10
2. Courtesy J. Lawson
3. Courtesy B. Janes; TNA AN 13/1376
4. Tonks, 1974
5. News Sheet No.1 (NGRS, May 1952)
6. D. Challinor, pers comm; R. Rowson, pers comm
7. Francis *et al*, 2000
8. M. J. L. Iggulden, personal communication
9. SA SC 3/1a 1/16
10. *London Gazette*, Iss.41107 p.28
11. SA SC 3/1a 1/17
12. A. Yapp, pers comm
13. NGRS; SDR library file
14. I. J. Brown, pers comm
15. SA SC 3/1a 1/18
16. Courtesy D. A. Brewer; SDR Sale Prospectus
17. *Narrow Gauge News* No.145 (NGRS, Dec 1984)
18. Land Registry

Chapter Ten

1. M. J. Messenger, pers comm; Messenger, M. J. *The Culm Valley Light Railway* (Truro: Twelveheads Press, 1993); Boyd, 1988
2. Cuckson, A. S. "Pontesbury Junction" IRR No.144, pp.103-119. Much concerning the construction and evolution of this station was previously published here.
3. TNA MT6/180/6; SA 4279/41
4. SA 800 43A
5. Author's notes, 1971
6. CSRMA; SDR Minute Book
7. SA 4743/48
8. R. G. Cooke, pers comm
9. Photographic evidence
10. Cuckson, A. S. "Snailbeach: Railway Staff, the repairs to DENNIS etc." IRR No.180 pp.89-98
11. Tonks, 1974
12. Cuckson, A. S. "Mining at Callow Hill" SCMC Journal No.8 (2003) pp.7-13
13. CSRMA, Hayward's Quarries Ltd plan; OS SJ 30 NE 1954 'Whole sheet revised pre 1930'
14. G. France, pers comm
15. V. J. Bradley, pers comm
16. NGRS; SDR library file
17. Boyd, J. I. C. "Snailbeach" (Glimpses of the Narrow Gauge No.8) *Railways* March 1951
18. SA 4743/50
19. J. Butler, pers comm
20. SA 5982 11/15
21. Cuckson, A. S. "The Snailbeach Barytes Mill" SCMC Journal No.7 (2001) pp.16-21
22. Courtesy N. Southwick, D. Boulter, pers comm
23. E. Rowson, pers comm
24. NLW; Longueville Collection, SDR Letter Books
25. Nell Rowson family collection
26. SA 4155
27. SA 5982 13/15
28. CSRMA; TNA BT 285/506
29. *Meccano Magazine* Vol.26 No.5 [May 1941] Letters p.170

Chapter Eleven

1. Woodcock, D. J. "Snailbeach locomotive shed" scale drawing (Narrow Lines No. 30, Dec 1984); survey of plan dimensions by triangulation by M. Bevan; sketches by D. A. Brewer; survey by the author 1971 and subsequent visits; Ove Arup and Partners, 'Snailbeach Lead Mine Study' Vol. 3 Appendices (England: Authors, 1985). Sketches B2.1- 4 showing structure and features of the locomotive shed have been a helpful reference but not always a correct depiction
2. SA 4743/48 Snailbeach District Railways working plan
3. Courtesy K. Lock; SA 5982 13/15 Mine Equipment Inventory & Valuation dated 1884 by A. Waters
4. Halton, P. S. "Baldwins in the Wood" *The Narrow Gauge* No.72 (England: Narrow Gauge Railway Society)
5. SA 5982 3/4
6. SA 800/43A

Chapter Twelve

1. SA 800/26C
2. R. Rowson, pers comm
3. R. G. Cooke, pers comm
4. CSRMA
5. TNA RAIL 1053 98
6. SA 800/43A
7. M. Rowson, pers comm
8. Nell Rowson family collection
9. NLW; Longueville Collection, SDR Letter Books
10. Cuckson, A. S. "Pontesbury Junction" IRR No.144 pp.103-119
11. G. France, pers comm
12. Courtesy M. J. Shaw; SA DA42/154/2/23
13. E. Rowson, pers comm
14. SA 4155
15. SA 800 43D
16. Courtesy M. J. Messenger; TNA MT6/331/9
17. SA 800/43B
18. Cuckson, A. S. "Snailbeach: Railway Staff, the repairs to DENNIS etc." IRR No.180 pp.89-98 (Mar 2005)
19. J. G. Vincent, pers comm

Chapter Thirteen

1. Cuckson, A. S. "BELMONT – the Snailbeach Hughes locomotive" IRR No.171 (December 2002), pp.232-239
2. SA 800/43A
3. Link, R. C. "Corris Conundrum" Parts 1 & 2 *NG Revw* Nos. 49 & 50 (RCL Publications)
4. Cuckson, A. S. "FERNHILL – the Snailbeach Barclay" IRR No.164 (March 2001), pp.463-466
5. Baker & Civil, 2008; A. C. Baker, pers comm
6. TNA RAIL 1053
7. NLW; Longueville Collection, SDR Letter Books
8. Tonks, 1974
9. Courtesy M. J. Shaw; SA 2478/80
10. CSRMA
11. Courtesy A. C. Baker; *The Locomotive Engineer's Pocket Book*
12. Baker, A. C. "DENNIS at Snailbeach" IRR No.164 (March 2001); Cuckson, A. S. "Snailbeach: Railway Staff, the repairs to DENNIS etc." IRR No.180 pp.89-98 (Mar 2005); CSRMA; SDR Minute Book

13. R. G. Cooke, pers comm
14. National Railway Museum; Kerr, Stuart locomotive register (copy) f.33
15. Gratton, 2005
16. Wells, 1962
17. Baker, A. C. "Bagnall and the WDLR 2ft 0in Gauge Locomotives" IRR No.130 pp.21-32 and subsequent letters
18. Courtesy Industrial Locomotive Society library; Notes from *Surplus*
19. R. C. Link, pers comm
20. Cuckson, A. S. "Snailbeach: the Baldwin Locomotives" *NG Revw* Nos. 56, 57 & 58 and subsequent letters (RCL Publications)
21. Boyd, 2007
22. P. Ingham, pers comm
23. *Railway Magazine* May-June 1944 p.183
24. Gratton & Band, 1989
25. SA 800/43B
26. Boyd, J. I. C. *The Talyllyn Railway* (Didcot: Wild Swan Publications Ltd, 1988); Incorporation details courtesy M. J. Messenger; TNA registers
27. M. J. Shaw, pers comm
28. SA 5982/128
29. SA 800/43C
30. Courtesy B. Janes; NRM PSH

Chapter Fourteen

1. SA 800/43A
2. TNA BT 31 2049/8975; Lancashire Record Office, Trade Directory
3. TNA RAIL 1053
4. NLW; Longueville Collection, SDR Letter Books
5. The Midland Railway Carriage & Wagon Co. (MRCWCo) has been suggested as the builder, but no proof has been found. Large collections of rolling stock drawings from the Metro-Cammell group (successors to the MRCWCo) are held at the Birmingham Libraries Archives and the Butterley headquarters of the Historical Model Railway Society, but enquiries have revealed no wagons bought for the SDR or the Granham's Moor Quarry Co. Ltd.
6. Tonks, 1974; A. C. Baker, pers comm
7. Wade, E. A. *The Plynlimon and Hafan Tramway* (Truro: Twelveheads Press, 2nd edition, 1997)
8. Courtesy R. C. Link; Notes by R. E. Tustin
9. TNA BT 285/506; "Notes and News" *Railway Magazine* May-June 1944 p.183, Sept-Oct 1944 p.310
10. Foley, P. letter in *NG Revw* No.59 p.131 (RCL Publications)
11. Cuckson, A. S. "Snailbeach Wagonery" *NG Revw* No.64 pp.335-341 (RCL Publications)
12. J. Butler, pers comm
13. CSRMA
14. G. France, pers comm
15. V. J. Bradley, pers comm
16. Courtesy M. J. Shaw; SA DA42/154/2/23
17. *Railway Magazine* Jan 1955 p.61
18. SA 5982 2/4
19. Dalston, H. F. G. "The Snailbeach District Railways" *Railway Magazine* Nov-Dec 1944 pp.355-358
20. Boyd, 2007

Chapter Fifteen

1. PA; Snailbeach Mine Estate Act, 1867
2. SA Longueville Collections 800 and 5982
3. Courtesy J. M. Taylor; genealogical research
4. SA 5982 2/15 SMCo Ltd Minute Book
5. SA 5982/34
6. CSRMA; SDR Minute Book; TNA BT 31/14804/21384
7. Lerry, G. G. "Henry Dennis: A Cornish Captain of Welsh Industry" Trans DHS 1 (1952), pp.33-45; Courtesy A. C. Baker; *Proceedings* (London; Institution of Civil Engineers)
8. Milner, 2008
9. Burt, 1977
10. SA 5982 2/15
11. SA 5982 2/15 SLMCo Ltd Minute Book
12. SA 5982 2/15; SA 2478/50-77
13. NLW; Longueville Collection, SDR Letter Books
14. P. Shaw, pers comm; M. J. L. Iggulden, pers comm
15. B. Janes, pers comm
16. TNA BT 285/506
17. SA; Census, Westbury
18. SA 5982 2/4
19. SA 5982 4/4
20. SA 5982 12/15; Courtesy I. J. Brown; *Minutes of Evidence – Commissioners in the Condition of Mines in Great Britain* Vol.7 Mining Accidents (British Parliamentary Papers, 1863)
21. SA 800/26C
22. Brown, I. J. "Captain Oldfield and the Shropshire Mines" *Below!* Iss.95.2, SCMC
23. SA 5982/128; SA 800/43B SDR pay sheets
24. P. Ingham, pers comm
25. Courtesy J. M. Taylor; census records
26. Courtesy J. M. Taylor; death certificate
27. W. J. Milner, pers comm
28. Milner, 1984
29. SA 4155
30. E. Humphreys, pers comm
31. CENSUS DETAILS OF SAMUEL JONES. One Samuel Jones has been found living near the SDR who might be

this long-serving employee. Born in Ruyton-XI-Towns, in 1871 he was a company agent (employer unknown) living next to the station master at Boycott on the Shrewsbury & Welshpool Railway – perhaps the right background for clerical and manual work at Pontesbury Junction. But in 1881 he was a mining labourer aged 46 living at Station Road, Pontesbury, and in 1891 was publican of the Horseshoe Inn. If this is the right man, he was also working during the day on the railway, as Daniel Simister did later. In 1901 he was a general labourer living at cottages known as The Wharf, Pontesbury, close to the SDR sidings. Nothing in the census records positively identifies him as a railway worker.

32. Courtesy J. M. Taylor; death certificate
33. CENSUS DETAILS OF THE HUMPHREYS FAMILY. Despite the spelling differences on the pay sheets, the J. Humphreys and George Humphries in the permanent way gang might have been brothers. A Humphreys family living at Station Yard, Pontesbury in 1881 are the only likely ones who might have been doing such work on the SDR – but living at that address, they might also have been moonlighting. John Humphries aged 19 was a railway platelayer, and his brother George aged 17 was a general labourer. In 1891 John was a railway labourer living at Pontesford. Nothing in the census records positively identifies them as working for the SDR.
34. Courtesy M. J. Shaw
35. CENSUS DETAILS OF WILLIAM ARTHUR JONES. It is unconfirmed, but Will Jones is probably the sawyer aged 24 living at Horsebridge, near Minsterley in 1901.
36. CSRMA; Mr & Mrs L. Hordley, pers comm
37. M. Rowson, pers comm
38. D. Harrison, pers comm; J. Butler, pers comm
39. "Potty characters at Kinnerley" *The Colonel* No.18 (Colonel Stephens Railway Society)
40. Strange, 1989; Redwood, 1981
41. Courtesy A. Austin; birth, marriage and death certificates; memorandum; notes by S. H. P. Higgins of interviews with former Bishop's Castle Railway staff
42. S. Rowson, pers comm
43. M. Chidley, pers comm
44. M. & D. Owen, pers comm
45. Dent, B. A. *The Snailbeach Methodist Chapel 1876-1976* Privately distributed paper
46. Weaver, R. "More about Snailbeach" IRR No.70 (April 1977)
47. R. G. Cooke, pers comm
48. N. Morris, pers comm

CHAPTER SIXTEEN

1. TNA RAIL 1053
2. TNA RAIL 1053 90
3. NLW; Longueville Collection, SDR Letter Books; TNA RAIL 1053 98
4. Brown, I. J. *A List of Fatal Accidents in Shropshire Mines from 1850-1979* Account 24 (SCMC, 2005)
5. SA 1950/4
6. Courtesy M. J. Shaw
7. M. Challinor, pers comm
8. J. Butler, pers comm
9. N. Morris, pers comm
10. M. Rowson, pers comm
11. R. Rowson, pers comm
12. Francis *et al*, 2000
13. R. G. Cooke, pers comm
14. Meccano Magazine Vol.26 No.5 p.170 (May 1941); letter from J. G. Vincent
15. V. J. Bradley, pers comm

CHAPTER SEVENTEEN

1. Courtesy I. J. Brown; *Minutes of Evidence – Commissioners in the Condition of Mines in Great Britain* Vol.7 Mining Accidents (British Parliamentary Papers, 1863)
2. SA 800 & 5982 collections, numerous instances; Burt, 1984
3. SA 800/25D
4. Brown, I. J. *Snailbeach Lead Mine* Account No.17 (SCMC, 1993)
5. SA 152/1
6. Martell, H. M. & Gill, M. C. "Lead Smelting in Welsh Furnaces at Pontesford, Shropshire" *Bulletin of the Peak District Mines Historical Society* Vol.11 No.6 (Winter 1992), pp.297-312
7. SA 5982 13/15
8. Courtesy J. Lawson; report of Thomas Sopwith
9. SA 5982 5/15
10. K. C. Lock, pers comm
11. SA 5982 2/15 SLMCo Ltd Minute Book
12. Brown, I. J. *A List of Fatal Accidents in Shropshire Mines from 1850-1979* Account 24 (SCMC, 2005)
13. SA 5982 10/15
14. Cuckson, A S. "Snailbeach Mine: pumping and winding machinery c.1782-1856" *British Mining* No.78 pp.118-130 (Sheffield: NMRS, 2005)
15. SA 800/26A
16. Trueman, M. R. G. & Gill, M. C. *Snailbeach Lead Mine Stage 2 Study* (Lancaster: Lancaster University Archaeological Unit, 1990)
17. OS County Series, Shropshire sheet XLVIII.6, 1881
18. SA 4155

19. SA 800/26C
20. Cuckson, A. S. "Mining at Callow Hill" SCMC Journal No.8 (2003) pp.7-13
21. Heathcote, J. "Geology of the South Shropshire Ore Deposits" *Account 22: Metal Mines of Shropshire*, p.48 (SCMC, 1997)
22. *Barytes and Witherite* 1922 British Geological Survey Vol. II
23. Courtesy M. J. Shaw; *Shrewsbury Chronicle* 11-7-1862 and 18-7-1862
24. SA 1950/4
25. PA; Snailbeach District Railways Act 1873; SA 4743/48
26. M. J. Shaw, pers comm
27. Nell Rowson family collection
28. Courtesy M. J. Shaw; J. M. Alldridge, pers comm
29. Hordley, J. *Turning all our Lead to Gold – An Account of the Burrs* (etc) (University of Birmingham dissertation, 2000)
30. SA 800/27A
31. SA 800/25C
32. SA 5982 12/15
33. http://www.minerals.usgs.gov
34. Shaw, M. J. The Shropshire Barytes Industry (University of Birmingham M.A. thesis, 2006)
35. http://www.galleries.com
36. http://www.mineralszone.com
37. http://www.geology.com
38. Burt et al, 1990
39. http://www.chemsoc.org
40. http://www.mii.org

Chapter Eighteen

1. Land Registry
2. Courtesy I. J. Brown; British Geological Survey Directory; Report of Dept. of the Environment, Transport and the Regions
3. Courtesy M. J. L. Iggulden; author unknown *Snailbeach: Addendum* a collection of notes post-1979 with incomplete reference to sources
4. CSRMA
5. G. Fairhurst, pers comm
6. D. Challinor, pers comm.
7. Web site: list.english-heritage.org.uk/resultsingle.aspx?uid=1017764 (3-5-2014)

Appendix One

1. Davies, 1966
2. John Milner Collection; GVT records
3. Courtesy W. J. Milner; Shropshire Archives (SA) 800/44B
4. SA 5982/128
5. Milner & Williams, 2011
6. SA 800/43B
7. TNA RAIL 1053

Appendix Two

1. SA 5982 1/4
2. SA 800/43A
3. Receipts with printed letterhead show that the weighbridge maker's name is Hodgson and *Stead*, not Head.
4. The numbers of wagons in the first two categories are miscounted according to the maker's advice note. See Chapter 13.

Appendix Four

1. Weaver, R. "More about Snailbeach" IRR No.70 (April 1977). Speculation in this article that the locomotive Dennis was bought because of a proposal for an extension line to Buxton can be dismissed on the evidence presented in Chapter 7
2. M. J. Shaw, pers comm
3. CSRMA
4. SA 3016/3
5. N. Rowson, pers comm
6. Courtesy M. J. Shaw; SA 4929/3/2

Appendix Five

1. All information courtesy M. J. Shaw
2. SA 731/5/11/153
3. SA 731/5/11/154
4. SA 731/5/11/152

INDEX

People